Quantum Physics for Beginners

Unraveling the Fundamentals of Quantum Mechanics, Particle Behavior, and the Nature of Reality along with the Interplay between Science and Spirituality

© Copyright 2024 - All rights reserved.

The content contained within this book may not be reproduced, duplicated, or transmitted without direct written permission from the author or the publisher.

Under no circumstances will any blame or legal responsibility be held against the publisher or author for any damages, reparation, or monetary loss due to the information contained within this book, either directly or indirectly.

Legal Notice:

This book is copyright-protected. It is only for personal use. You cannot amend, distribute, sell, use, quote, or paraphrase any part of the content within this book without the consent of the author or publisher.

Disclaimer Notice:

Please note the information contained within this document is for educational and entertainment purposes only. All effort has been executed to present accurate, up-to-date, reliable, and complete information. No warranties of any kind are declared or implied. Readers acknowledge that the author is not engaging in the rendering of legal, financial, medical, or professional advice. The content within this book has been derived from various sources. Please consult a licensed professional before attempting any techniques outlined in this book.

By reading this document, the reader agrees that under no circumstances is the author responsible for any losses, direct or indirect, that are incurred as a result of the use of the information contained within this document, including, but not limited to, errors, omissions, or inaccuracies.

Your Free Gift
(only available for a limited time)

Thanks for getting this book! If you want to learn more about various spirituality topics, then join Mari Silva's community and get a free guided meditation MP3 for awakening your third eye. This guided meditation mp3 is designed to open and strengthen ones third eye so you can experience a higher state of consciousness. Simply visit the link below the image to get started.

https://spiritualityspot.com/meditation

Or, Scan the QR code!

Table of Contents

INTRODUCTION 1
CHAPTER 1: INTRODUCTION TO QUANTUM PHYSICS 3
CHAPTER 2: EXPLORING PARTICLE BEHAVIOR 15
CHAPTER 3: WHAT IS LIGHT? 24
CHAPTER 4: QUANTUM OBSERVATIONS, EXPERIMENTS, AND THEIR INTERPRETATIONS 35
CHAPTER 5: QUANTUM REALITY AND CONSCIOUSNESS 48
CHAPTER 6: QUANTUM MYSTICISM - SCIENCE AND SPIRITUALITY ... 62
CHAPTER 7: ENTANGLEMENT - EVERYTHING IS CONNECTED 74
CHAPTER 8: SUPERPOSITION: ANYTHING IS POSSIBLE 84
CHAPTER 9: THE MULTIVERSE 93
CONCLUSION 102
HERE'S ANOTHER BOOK BY MARI SILVA THAT YOU MIGHT LIKE 104
YOUR FREE GIFT (ONLY AVAILABLE FOR A LIMITED TIME) 105
REFERENCES 106

Introduction

If you know next to nothing about physics, let alone quantum physics, this topic may initially appear intimidating. However, with this book in your hands, you won't feel that way for much longer. If you've always wanted to know what the secrets of the subatomic world are, and you want to dig up the mysteries shrouded in scientific jargon, you couldn't have picked a better book for the job than this. No scientific background? No problem. Every complex concept of quantum physics is broken down into language that is easy to understand. You won't have to grapple with complex formulas or try to work out complex equations, either.

Unlike other books on the topic of quantum physics, this book is suitable for beginners. The concepts are clear and explained in engaging language. It is a book full of carefully deconstructed concepts to help you understand the incredibly quirky world of quantum physics.

It offers so much more to readers than just an understanding of this branch of science. This book is excellent for those who've always sought the bridge between spirituality and science. It is full of practical information demonstrating how to improve your life. By the final page, you'll discover that your understanding of reality is deeper and richer than it has ever been.

It is no accident you've chosen this book from all the others you could be reading right now. You've been handed a passport to the power to unlock every world you could think of. If you boldly choose to continue reading, rest assured that your life will no longer remain the same. Therefore, you should exercise caution and only dive into this book if you are prepared to have your socks knocked off.

You will learn everything you need to know about reality, consciousness, and your specific purpose in life. This is no ordinary textbook. This is a whole experience, one you're not about to forget. So, if you are prepared for the new, the strange, and the extraordinary, there's not a moment left to lose. Begin with the first chapter and discover the magic in reality.

Chapter 1: Introduction to Quantum Physics

Suppose you know nothing about the sciences like physics. In that case, quantum physics may seem like such an intimidating topic that you'd never normally touch it with a ten-foot pole. As you'll soon discover, it's not that difficult to understand. This chapter will introduce you to quantum physics, keeping things as beginner-friendly as possible. You don't have to worry about pulling your hair out to understand the concepts. That's a guarantee.

Welcome to Quantum Physics. An electron around a nucleus.
Designed by Freepik. https://img.freepik.com/free-photo/atom-science-biotechnology-blue-neon-graphic_53876-167297.jpg?t=st=1712095432~exp=1712099032~hmac=56d0a39abad98fe04895eb12e59f75303 0c1892186665aa0b00ec0a86a17b798&w=1060

The Puzzle of Quantum Physics

Hello, Alice. Sure, that's not your actual name, but it might as well be because you're about to discover how deep the rabbit hole goes and how crazy things get in Wonderland. This sentiment may appear an exaggeration, but slowly, you'll find it's anything but. The quantum universe really is that illogical. The rules of reality are anything but what they seem regarding quantum physics. Everything about the nature of reality will bend, boggle, befuddle, and blow your mind once you discover what quantum physics is about.

Imagine finding out your body is in two places or more at the same time, and that one version of yourself is over at Buckingham Palace racking your brain about how to remain relevant in this day and age, while another version of you is sipping a mai tai somewhere in Bali.

Picture flicking the light switch in your room on and off, and each time the light's on, your room is a different version of itself. Your wall paint, bed position, stuffed animals you're embarrassed to admit you still have, and the bazillion pillows are different every time. It sounds chaotic, doesn't it?

You know those socks of yours that mysteriously go missing? What if one of a pair is in Pluto and the other with you? Also, what if every time a Plutonian washes or dirties that sock, you can tell – *because your sock is a reflection of the alien's?* What does all this have to do with quantum physics?

First, you have to know the difference between classical physics and quantum physics. Classical physics is everything that pertains to the rules of the physical world as you perceive it with your five senses. It's stable and predictable. You know that if you throw a basketball in the air, it will return to Earth. Throw the ball hard against the floor, and it bounces. Pull a door open, and it moves toward you. Try to sit your ample bottom on a toddler's chair; with enough time and weight, it will break. Your bank is always at the same address, never moving from that spot, and the speed of light is fixed.

The laws of classical physics are fixed, unbreakable, and dependable, which is nice because how weird would it be to discover the sun now rises in the north or that the chair you're looking at in the corner isn't actually there right now? To put it formally, classical physics is a set of theoretical perspectives meant to explain observable phenomena and objects, such as planets, sound, light, cars, etc. The field of science studies the why and how of the movement of things, as well as how they work, unpacking the mechanics of magnetism, electricity, motion, sound, heat, gravity, and light.

Now you understand the basics of the classical version of physics, what about its quantum counterpart? Based on the introduction to the concept already offered, you're probably assuming it's nothing but the wild imagination of a fictional character in a cartoon – something cooked up in Dexter's Laboratory, perhaps. You'd be forgiven for thinking so. If quantum physics were a clock, the next minute after 7 AM would be 33:56 PM.

In other words, nothing quite lines up with the rules of classical physics. Depending on who's looking at it, one object is multiple things, all simultaneously. In this world, the speed of light isn't the fastest thing.

Okay, so you get the point and want a straightforward definition of quantum physics. Basically, **quantum physics is the field of science that studies the fundamentals of matter and energy, seeking to explain the universe at the level of atoms, electrons, and photons.**

The fact that you're reading this book means you've likely heard of quantum mechanics, too. Put simply, quantum mechanics is the mathematical language that describes the way atomic and subatomic particles move and interact with one another, working within frameworks like the uncertainty principle, correspondence principle, wave-particle duality, etc. Don't worry; these things will sound less like gibberish as you explore this quantum world further. In some contexts, quantum physics and quantum mechanics are used interchangeably.

Now, you could use a history lesson.

Historical Context and Development of Quantum Theory

Before getting into anything else, credit must go where it's due. Max Planck was the one who came up with the quantum theory. Without him, the many other fascinating discoveries in this field may have remained forever unknown. This German physicist published a study that sent shockwaves through his field.

Max Planck came up with the quantum theory.
https://picryl.com/media/max-planck-1933-1bf0ff

The study was about how radiation affects a "blackbody" substance, which is something that absorbs all light and energy with which it comes in contact. He found that there are times when energy acts like physical matter. According to classical physics, energy is only ever in the form of a wave. However, Planck had a theory that these waves actually had particles he dubbed *quanta*. He won the Nobel Prize for his groundbreaking work.

Albert Einstein built upon Planck's work. First, in 1905, he posited that light is actually made of particles. This was dismissed as preposterous at the time because everyone assumed that light was in *waveform*. He called light particles "photons" and stated that each one has energy within it.

Then, four years later, in 1909, Einstein shook the science world again with his wave-particle theory, stating that waves and particles can behave like one another, especially regarding electrons and photons. Why was this such a big deal? Well, if you read between the lines, he was essentially saying that a particle can act like a wave and a wave like a particle, *depending on how you look at them.* This was just one of several theories Einstein proved, even though he wasn't entirely a proponent of quantum mechanics, as he didn't like the idea of an uncertain reality. In his words, "God does not play dice."

Sidenote: Depending on whom you ask, Einstein actually got all his ideas from Mileva Marić, his wife, but because this was a time when women weren't acknowledged as much as they should have been for their brilliant minds, he got all the credit. This book isn't about that debate, so it's time to move on.

In 1913, Niels Bohr used the quantum concept to explain the structure of atoms and molecules. In his model, the nucleus is at the center of the atom, much like the sun is in the center of the known planets. The electrons are set up like planets around the nucleus, but their orbit doesn't stray beyond specific distances (called "energy levels") from their "sun."

In 1924, Louis de Broglie contributed to the quantum field of study by taking Einstein's original position even further. To Louis, light wasn't the only thing with the traits of waves and particles. He ascribed that property to everything in existence. In other words, simultaneously, everything can be a ball or an ocean wave. One must wonder what Einstein would have thought about that.

Werner Heisenberg would not only devise a different way to work out the math of quantum mechanics within the context of matrices but also introduce the world to his "Uncertainty Principle" in 1925. He might as well have said to Einstein, "God, in fact, *does* play dice."

There's no better analogy for his theory than a hummingbird in flight. Watch its wings, and you'll notice only one of two things: the speed at which the bird's wings beat against the air or the wings in a specific spot. You can only pick up on one or the other, not both simultaneously. This is a rather simplistic example, but it explains the weirdness of the Uncertainty Principle.

Of course, quantum physics wouldn't be what it is without the work of that one scientist who may or may not have had a cat at a certain point in his life. His name? Erwin Schrödinger. His Wave Theory of Matter validates Niels Bohr's insistence that God may fancy a game of roulette now and then. His eponymously named Schrödinger's Equation, formulated in 1926, offers a mathematical way to describe how the quantum state of a quantum system evolves over time.

Solutions derived from Erwin's equation offer an excellent way to tell what the probabilities of various outcomes may be, demonstrating clearly that a particle can exist in more states than you can count until you actually observe it, which fixes it to a single state – at least, while you're looking. If you're unfamiliar with Erwin's contributions, surely you know about Schrödinger's cat thought experiment. If you aren't, you will be soon enough!

Paul Dirac was another interesting character who took Einstein's Relativity Theory to a quantum level. You see, this particular theory of Einstein says that while the laws of physics work the same for everyone all over the world, what you observe may be different from what someone else observes, depending on the direction of the movement of an object and its speed.

So, Dirac took this theory and applied it to the quantum world. He developed the Dirac Equation, which demonstrated how electrons and similar particles act whenever their movement approaches the speed of light. The man was able to predict that antimatter was a real thing. What's that? Antimatter is the mirrored version of matter, having the *opposite energetic charge.*

In 1932, Carl D. Anderson validated Paul's assumption that antimatter exists, thanks to his discovery of the positron as he was looking into the behavior of particles with a high energy charge known as cosmic rays from space. As he tracked the particles using his cloud chamber device, he noticed certain tracks left behind by said particles that appeared to be similar in mass to the electron – but positively charged instead. When he experimented with shooting high-energy light or gamma rays into various materials, he proved conclusively that each electron is paired with a positron.

One more honorable mention is Richard Feynman, who did phenomenal work on Quantum Electrodynamics or QED, which offers clarity on the interaction between electrons (matter) and photons (light). He created the Feynman diagrams, which acted as roadmaps to show how this interaction evolves over time, making handling all the complex calculations in QED much easier.

Richard Feynman.
https://picryl.com/media/richard-feynman-1988-2d6dca

Not only that, but the brilliant Feynman devised a simple way to work out all the possible ways a particle can journey from one point to another. It's like knowing every possible path that the ants that have been bothering you at home could make, from their nest to the cake sitting on your kitchen counter, including the illogical paths. Feynman's work also made quantum computing and nanotechnology possible, and he took it upon himself to teach physics principles to laypeople, breaking complex ideas into simple forms, much like this book does.

The Cornerstones of Quantum Physics

So . . . the bell has rung, and history class is over. Now, it's time to explore the different theories that, together, make up the fabric of quantum physics. Don't worry; you'll get clear explanations that won't make you want to gouge your eyes out.

Quantum Field Theory: Also called QFT, this theory combines quantum mechanics principles, which govern the nature of subatomic particles, with relativity, which is about everything to do with large distances and high speeds. Thanks to QFT, everyone now understands how subatomic particles interact through various force fields. Think of the world as being a giant ocean of energy. Particles such as photons and electrons act as ripples or waves in this universal energetic ocean. What's causing the ripples? Energy itself.

So, according to the Quantum Field Theory, the particles act as excitations that cause rippling waves that occur in their underlying fields. What are these underlying fields? Think of them as energetic blankets all over the universe. There is a separate blanket (field) for each kind of particle. When you pick a point in a field and add energy to it, the energy causes a disturbance or ripple, which is the particle itself.

String Theory: According to this theoretical framework, particles aren't simply infinitesimally small points or dots in space as depicted in regular physics, but more like the tiniest pieces of string. The strings can vibrate, and how they vibrate determines the particle's mass, energetic charge, and other unique traits. So, this theory looks into how these strings travel through space and how they affect one another in the process.

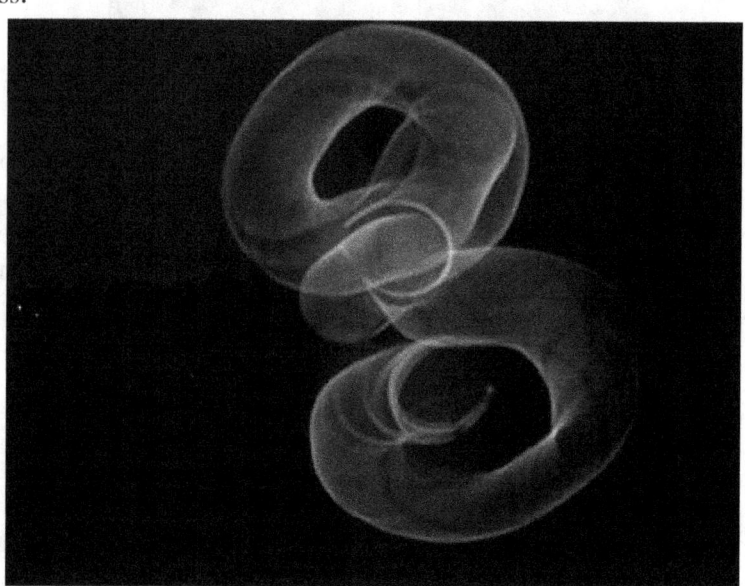

One of many visual interpretations of string theory.
https://www.flickr.com/photos/7725050@N06/631503428

There are various vibrational states these strings can be in, one of the most important ones being the graviton. The graviton is a particle governed by quantum mechanics that contains the gravitational force, and this is why string theory is also known as quantum gravity theory. It is an all-encompassing theory that explains everything in the universe using the language of mathematics, describing every force in existence and matter in all its forms, known and unknown.

One implication of this theory is that there are other universes besides the one you know, which operate by different laws of physics, and that there are other dimensions beyond what is known about time and space that remain imperceptible – for now. If you think that's crazy, try this on for size: If string theory is correct, that would mean the universe is actually a hologram. Is a sense of existential crisis looming in your mind? You may want to hold off on that for now.

Wave-Particle Duality: This concept suggests that all particles can act as both waves and particles. Think of light. Focus it on a surface, and its photoelectric effect can knock the object's electrons loose, which shows that light can act like a particle. It is like when you're playing pool, and the white cue ball strikes another ball to make it move. The other ball moves because the white ball transfers energy to it once it makes contact, and, in the context of light hitting a surface, it also causes movement by knocking electrons out of place.

If you take that same light and let it shine through a narrow slit, guess what? It acts like a wave because it will cause an interference pattern. What's that? It's the light forming a pattern of light and dark bands, which is something waves do. You can understand this by thinking of the ripples of waves created when you drop a pebble into a pond. Light can be wavy, too. It's the same thing for other forms of matter, like electrons. As particles, they can move from one specific spot to another, but as waves, they spread out, so they aren't tied down to one location but are in multiple places at once.

Quantum Superposition: This is a quantum mechanics concept that states all particles exist in more than one state simultaneously unless and until someone observes them. Remember Schrödinger's cat? Well, the science behind that is that it's attention and observation that fixes the position and state of a particle.

Particles are in a superposition, which means multiple positions. They're not moving around these positions but are in every one. Have you ever heard your kooky New Age friend say something like, "There's only here and now?" Well, this concept is the scientific way to explain that. It suggests that particles act like all that exists is here and now, at least until you pay some attention to them! When you do, they'll pick a spot.

Quantum Entanglement: This quantum mechanics concept is one where a pair of particles become connected to each other, so even if they're the furthest they could be from each other, whatever changes one of them experiences will be reflected in the other – and that's why your nonexistent sock from earlier in this book keeps washing and dirtying itself.

To put it in scientific, non-sock terms, when you know the measurements of a particle, you know the same of the other particle. Now, you may think that surely, at some point in space beyond a certain distance, the connection between these particles must be broken. After all, isn't that kind of how WiFi works? Head outside and walk far enough from the house, and you'll lose the connection to the network at home, right?

Well, that's not quite it.

These linked particles could be lightyears apart, yet they'll still mirror each other because they're entangled. Einstein referred to this as "spooky action at a distance." That's a fitting description of quantum entanglement, and if you think about it, that explains why certain spiritual practices require "sympathetic magic," where practitioners like Voodooists, for instance, use items to represent the people they'd like to help or hex.

The Uncertainty Principle: Also called Heisenberg's Uncertainty Principle, this concept states there's no way to be aware of a particle's location and precise speed as it moves in a specific direction *simultaneously*. You can only know one or the other. When you can track the location precisely, you won't be able to do the same for its speed, and vice versa. What gives? Are scientific instruments pointless? No, that's not the case. This is simply the way quantum particles operate. It's as if their principle is the meme, "Never let them know your next move."

Quantum Tunneling: If you throw a ball against the wall, you expect the wall to stop it in its tracks, right? Also, if you roll that same ball down a hill, you expect it to keep going, right? In quantum physics, there is a concept called quantum tunneling, which would suggest that rather than the ball being stopped by the wall or going over the hill, it would pass through both obstacles.

So, you can see that this theory would not work in classical physics because if you attempt to drive a car through a barrier like a gate, it will lead to a terrible accident. However, quantum tunneling occurs frequently in quantum physics, as particles move or "tunnel" through obstacles or barriers, like a hot knife through nonexistent butter.

Quantum Physics, Applied

There are numerous ways quantum physics can be applied in modern technology. Here's a quick overview of some of them. First, consider lasers. These work using stimulated emission. In simple English, a light particle (photon) is used to cause a reaction or "stimulate" an already excited electron, causing its energy state to drop lower, resulting in the release of two photons that are similar in nature, giving you a powerful, concentrated light beam. This entire process relies on quantum physics.

What about transistors? Modern electronics, as you know them, wouldn't exist without them, which would be a bummer. The transistors are the cornerstone of all electronics, and they work with the principle of quantum mechanics, which is how electricity flows where it should through circuits.

Even the world of medicine benefits from quantum physics. Magnetic Resonance Imaging (or MRI) machines are necessary to diagnose the problems patients deal with since they offer a clear snapshot of what's going on inside the body. Magnetic Resonance Imaging works with quantum physics. How? This imaging technique is made possible by controlling how atomic nuclei spin and picking up on the resulting radio waves when the nuclei go back to their actual state.

Then there's cryptography, which is necessary to ensure no third party can decipher messages sent end-to-end. If you get a message, you're the only one who can read it, and no one else can. There's research going on in the security field that's working to incorporate quantum principles into the cryptography process. How?

Remember the Uncertainty Principle? If someone observes a particle, its behavior will change. In quantum cryptography, if an eavesdropper is trying to get in between you and your message when it's transmitted to you, the original message will be altered significantly, and this alerts you and the sender to the fact that one of you has a personal FBI agent assigned to them.

Things to Remember

1. In quantum physics, energy isn't some nebulous thing that cannot be measured. It occurs in discrete units.
2. Particles also act as waves.
3. Particles exist in several states at the same time.
4. Two particles can become linked, mirroring each other no matter their distance.
5. There's no way to tell the speed and location of a particle with precision.
6. Particles can move through various potential energy constraints, which is something classical physics insists is an impossibility.
7. Schrödinger's cat is really an explanation of superposition and doesn't mean your cat dies every time you can't find it.

Chapter 2: Exploring Particle Behavior

Now you've been introduced to the wacky world of quantum physics, what's next? In this chapter, it is time to dive into the behavior of particles, both as waves and particles on the quantum level.

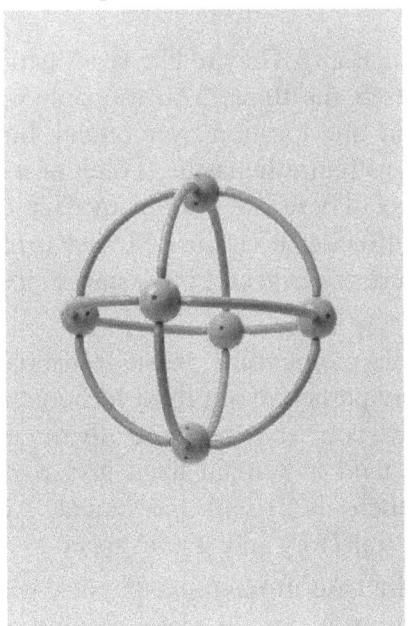

Particles on the quantum level.
Designed by Freepik. https://www.freepik.com/free-photo/chemical-element-arrangement-still-life_16691170.htm#fromView=search&page=2&position=16&uuid=cbd7f1b0-2c6a-4ea7-84db-dbb5908fa2d9

Classical Particles Versus Quantum Particles

If you're going to get this quantum physics thing down, it's essential to know the difference between the way particles behave in the quantum framework versus the classical one. You already know some of this based on what you've learned from the previous chapter, but it doesn't hurt to go over them once more to be thorough.

Deterministic behavior versus probabilistic behavior: When it comes to classical physics, you can determine how a particle is going to act in the future once you understand its current state. That's why it's known as deterministic. There are rules, and all the rules are respected. When you know the speed at which a planet moves and its present position, you can work out where it will be at a later date with little to no wiggle room for error. Classical physics has to do with macroscopic objects – as in objects large enough for you to observe with the *naked eye*.

On the flip side, quantum physics involves microscopic objects, and in this world, probability reigns supreme because nothing is definite. What's 1 plus 1? It's probably 2, 20, or a zebra. That's because quantum particles live in a world with multiple possibilities that are all dependent on wave functions.

What are wave functions? They're like road maps that tell you how to find the particle. Here's the thing. You can only work out the odds of finding the particle in one location over others, but that doesn't mean you've pinned down its future behavior. Think of it like trying to predict where smoke will go. There are too many factors to track, so your prediction will be imprecise. If classical physics is black and white, then quantum physics is every shade of gray – and every other color known and unknown.

Fixed location or momentum versus uncertainty: According to classical physics, every particle has a fixed location and travels at a fixed speed. This is why you can measure them precisely. Generally, macroscopic objects tend to remain fixed in a particular position, and when they aren't, then they're in movement, which means they're traveling in a specific trajectory and at a set speed.

The opposite is the case in quantum physics, where every particle is potentially a wave as well. So, in classical physics, you'd think of a particle as a ball, while in quantum physics, it would be more like a wave or a plume of smoke. There is no way you could pick out "one smoke" from that plume, and if you could, that would be an awesome trick to see!

One reason trying to measure both the position and speed of a particle in quantum physics is a fool's errand is because the very act of observing it or attempting to measure it would cause a disturbance in its momentum and vice versa.

Single state versus superposition: You already know what superposition is, so there's no need to bore you with the details. However, it is important to know that superposition is a concept that only applies to quantum physics and not classical physics. In classical physics, a particle can only be in one location at any given time.

To help you understand this better, imagine that you have a coin. The coin should have two sides to it: heads and tails. If you flip that coin and it lands, it can only land on heads or tails. Sure, you could probably pull off a trick shot and let it land on its side, but that's not what this is about, so don't be cheeky. Even if it did land on its side, the point is in classical physics, your coin would be on its side only, not on its side, heads, and tails, as would be the case in quantum physics.

Independence versus entanglement: Now you have two coins (you're a real Mr. Moneybags, aren't you?) Flip both coins simultaneously, and the odds are that they *will* land. However, they will land independently of each other. Whether they both land on heads or tails or land on different sides, it all comes down to the starting conditions before you flipped the coins and other factors that may have affected them as they turned in the air and landed on the ground. This is how it works in classical physics. Neither coin affects the other. Every particle, according to classical physics, is independent.

However, in quantum physics, those coins can develop a connection, or a sort of "energetic chemistry," if you will. Whether your coins are in the same room with you or galaxies apart, they affect each other. So if you're getting heads when you flip your coin, whoever has the other coin is also getting heads. In quantum physics, these coins (particles) are entangled. Thanks to Einstein's "spooky action at a distance," both coins are connected with each other and can communicate with each other.

Now, granted, the example with the coins is a little bit simplistic because the thing about particles in quantum physics is that while the entangled particles' fates affect one another, it is impossible to predict what will happen to them in the future after or while attempting to measure them.

Continuous energy spectrum versus discrete energy levels: Think about how you can gradually increase or decrease the volume of your TV, or think about a dimmer switch and how you can gradually reduce or increase the brightness in your room. You see, according to classical physics, you can add or remove energy in infinitesimally small amounts to cause smooth motion.

According to classical physics, there is no limit to the amount of energy that can be contained within a particle. However, according to quantum physics, each particle has discrete energy levels which are distinct and measurable. What does this mean? As you switch from one level of energy to another, in classical physics, the movement is smooth and continuous. That's not the case in quantum physics because each particle has a discrete energy level. It jumps from one level to another.

Now, come back to the dimmer switch analogy. The particle can go straight from light to darkness and back. The dimmer switch wouldn't be a *dimmer switch* as the particles teleport from dark to light right away rather than through a gradual increase of light. This is why neon signs have a vibrant glow as you watch the electrons teleport around to create it.

Diving Deeper into Wave-Particle Duality

While you've already been introduced to the concept of wave-particle duality, there's much more to explore. In 1928, Niels Bohr developed his complementarity principle, which posits that the only way to truly understand quantum phenomena is by being fully conversant with the properties of waves and particles. You could set up an experiment that would cause photons and electrons to act like waves. Make a few tweaks to your setup, and the next thing you know, these particles are acting like particles instead.

Niels Bohr.
https://pixel17.com, CC BY-SA 2.0 <https://creativecommons.org/licenses/by-sa/2.0>, via Wikimedia Commons
https://upload.wikimedia.org/wikipedia/commons/9/9a/Niels_Bohr_Portrait.jpg

But how can you tell the difference between the two? When a quantum particle is being a particle, it can dislodge electrons from surfaces. You see this play out in the photoelectric effect, which was discovered by your favorite white-haired genius, Albert Einstein, in 1905. So here's a breakdown. Think of light as a continuous flowing wave, much like the ripples in a lake or a pond.

According to classical physics, when there is a gradual increase in the brightness or wave intensity, you expect there to be a corresponding gradual increase in the energy that is transferred onto the electrons of, say, a metal surface. However, it wouldn't matter how bright the light gets. That wouldn't be enough to cause electrons to be ejected from the surface.

After a few tests, Einstein realized that no matter how strong the light is, it will always eject electrons from a metal surface if the energy frequency is above a certain threshold. Put differently, if you had the brightest light and its frequency wasn't high enough, the electrons would remain in place.

However, if the light is weak, it can cause electron emission if its frequency is high. The only particles capable of dislodging electrons are the photons that have energy beyond the metal's work function, which is the least amount of energy you'd need to trigger electron emission.

On the flip side, when photons and electrons act like waves, they can interfere with one another, and it's this interference that leads to light patterns and dark bands reminiscent of a rippling pond.

Now, bring your attention back to the complementarity principle. This principle makes it clear that you cannot observe the particle traits and wave traits of a particle at the same time. Nevertheless, you have to consider both of them simultaneously to be able to describe them fully since they complement each other. Wave-particle duality is a particularly useful concept when it comes to photonics, electron microscopy, quantum computing, and semiconductor devices, among other applications.

More on Superposition

As you've discovered, particles in quantum physics are never in a definite state unless and until you observe them. For instance, if you like to play the lottery, it's like having a ticket and not scratching it yet. Until the point when you scratch it, it remains both a winning and a losing ticket. A particle like an electron is here *and* there, but when you finally observe it, it chooses to be here *or* there. This isn't witchcraft. This is a concept that has been proven with laboratory experiments using electrons. You'll learn more about a famous experiment called the *double-slit experiment* in a later chapter.

The concept of superposition can also be found in quantum computing. A bit is the smallest unit of information that is used in computing. When it comes to quantum computing, the bits are called quantum bits or qubits. In this context, the qubit can be in both 0 and 1 states (remember, computers work with the binary of 0 and 1).

It is this ability that makes it possible for a quantum computer to outperform regular computers, as they're able to solve the most complex calculations you could imagine at record-breaking speed. The minute you give attention to a quantum system, it will be forced to pick one of the possible states available. This process is called a **wave function collapse.** Once it fixes itself to a state, every other possibility becomes nonexistent.

Superposition completely rips apart the ideas of determinism and locality, cornerstones of classical physics. This is one of the reasons that classical physicists are opposed to quantum physics. Who could blame them? After all, it is a little scary to think the universe's future cannot be predicted. If there's one thing most people fear, it's the unknown. Also, can you imagine a world where nothing is fixed in place?

Imagine inviting a friend over for brunch. They ask you when they should show up, and you tell them to come by 12:00 PM two years from now or by 5:00 PM three weeks ago.

They ask you for your address, and you tell them it's *probably* on the corner of Diagon Alley and 6th Avenue and probably in Sector 12 on the dark side of the moon. They'll need all the luck they can get to find an Uber to get them there, and if Rolex can figure out how to create a quantum wristwatch, they'd make a killing!

All this is to say that when it comes to superposition, the definite does not exist. Without the observer effect (the phenomenon where your attention on a particle fixes it in space and time), it's all in a haze of probabilities.

Into the Quantum Tunnel

The theory of quantum tunneling has already been explained in the previous chapter. In classical physics, when you throw a baseball against the wall, it will bounce back. That ball can't pass through the wall unless, of course, it's the platform 9 ¾ in the "Potterverse." It's also impossible for that ball to make its way over the hill unless you put enough kinetic energy behind it to get it moving.

When it comes to quantum physics, quantum tunneling is a thing. The tunneling particle doesn't need "enough" energy to get through obstacles. How is this even possible? In quantum physics, you talk about particles in the context of wave functions, which are mathematical functions that explain the probability of locating a specific particle in various locations.

Every time a particle comes face to face with an obstacle or barrier, the wave function will go down dramatically, but never so low as to hit zero while it is within the obstacle. Since there's a nonzero probability of finding the particle on the opposite side of the barrier, this is what allows the particle to make a tunnel through it.

Is this starting to sound a little too much like gibberish? Well put simply, nonzero probability means a "tiny chance." The wave function is basically a cloud or a plume of smoke that surrounds the particle. If you have a particularly dense cloud, you are likely to find the particle within it.

A dramatic decrease in the wave function means that the cloud would become thinner and thinner as you move further into the barrier that obstructs the particle. The thinner that cloud is, the less likely you will find the particle in it. Now, just because the wave function has decreased doesn't mean it totally disappears when it's within the barrier or obstacle, which gives the particle that "slim chance" or nonzero probability of popping up on the other side.

Particle Collisions

Every particle has its unique field. The fields make it possible for particles to connect with one another. Look at electrons, for instance. Their interaction involves the exchange of virtual photons, which are different from regular ones. Regular photons are the particles you can pick up on as electromagnetic radiation or light, while virtual photons are mathematical tools you use in quantum physics to explain how charged particles interact with one another. They're "virtual" because there's no way to directly observe them, and they are used to hold all sorts of energies – including energies that aren't physical.

There's no way to understand the quantum world fully without knowing about particle collision, which is a term to explain the way particles interact. These collisions last a short period and could be between subatomic particles such as protons and electrons- or larger ones like molecules and atoms.

There are three kinds of collisions: elastic, inelastic, and perfectly inelastic. In elastic collisions, the speed (momentum) and kinetic energy (the energy required to make the particles move) are conserved in the process. The inelastic collisions will conserve the momentum but not the kinetic energy. As for the perfectly inelastic collisions, after the particles have collided, they remain stuck together.

Scientists look at these interactions to better understand the laws of the quantum world, and they work with particle accelerators like the Large Hadron Collider at CERN to learn more. Devices like these basically cause the particles to move at high speeds and then bump up against one another with enough force to form new particles that can be studied. This process is how the Higgs boson particle was discovered. This particle is unique because it's responsible for giving mass to other particles.

The "Okay, Buts" of Quantum Physics

So, you understand the basics of quantum physics, but what gives? Surely, certain concepts won't fit into the box of quantum physics, right? Well, you're correct in that assumption. To wrap up this chapter, here are some of the challenges with this branch of physics:

1. The fact that quantum mechanics and physics are about studying the microscopic universe makes it difficult for many to relate to, and it certainly doesn't help that many of the discoveries and theories in this field do not fit in with common sense.
2. Particular aspects of quantum physics are still quite a challenge to apply in the real, macroscopic world, especially when it comes to technology.
3. In practice, superposition eventually breaks down into a state of decoherence where the particle that was "everything, everywhere, all at once" is now just the one thing in one state in the present. Superposition could be a boon to quantum computing as soon as some genius figures out how to halt the decoherence process.

Still Elusive

A few aspects of particle behavior are still under review in quantum physics. For instance, some particles have memory, as they're able to keep track of their pasts. These are called *non-abelian anyons*, and research on them has been going on for decades now. Then, there's another set of particles known as *neutrinos*. As they travel through space, they can change from one type or "flavor" of a particle to another, oscillating from one of three flavors to another as they move. They could be electron neutrinos, muon neutrinos, or tau neutrinos. Finally, certain metals are anything but conventional, having high-energy particles that may hold some promise in helping scientists find a new way to craft detectors that can pick up on wavelengths scientific instruments can't detect right now.

Chapter 3: What is Light?

Now, it's time to talk about light. You may not realize it, but there's so much more to light than just "light," you know? You'll learn more about it in this chapter as you discover the quantum theory of light, which explains how it behaves when you study it from a quantum level.

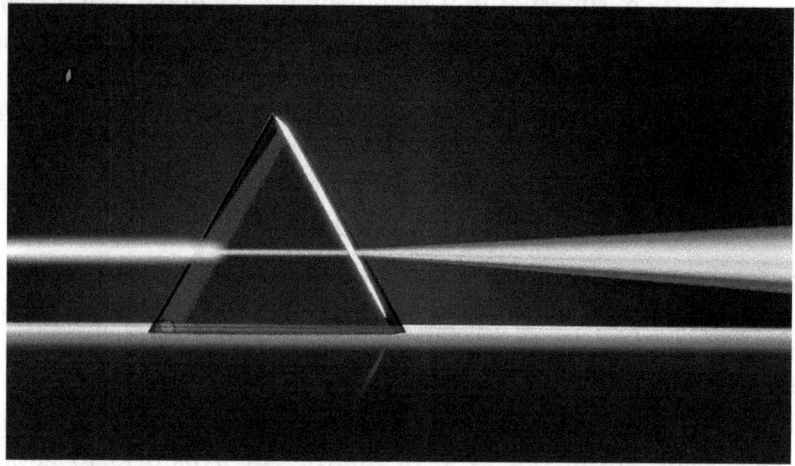

Discover the quantum theory of light.
https://pixabay.com/photos/prism-light-spectrum-optics-6174502/

From Classical Optics to Quantum Optics

You can't talk about light without talking about optics. But what do "optics" mean in the first place? Classical optics is an aspect of physics that seeks to offer a comprehensive description of light, with the perspective that light rays move through space in straight lines through objects that are large enough for you to see with your eyes with no special instrument.

In this branch of optics, scientists are curious to know how light makes its way through glass, water, air, and other macroscopic media like those. Every time light passes through these media, it doesn't do so quietly. If you could observe the molecules and atoms of the medium, you'd notice the light doing a lot of magic. For instance, it bends when it has to go from one medium to another through the process of refraction. Sometimes, it bounces to create a reflection; other times, it's absorbed by the medium.

What about quantum optics? This is an aspect of AMO physics, Atomic, Molecular, and Optical Physics, which is about how light interacts with quantum matter. You're definitely going to need special instruments to observe light in action because human eyes can't pick up on quantum information – at least not yet. Perhaps that's something that will change with newer iterations of Elon Musk's Neuralink.

Quantum physicists look at light differently. Ask one, and they'll tell you it's made of photons, which are discrete packets of energy. What do they mean by "discrete," though? Is it a "secret light?" You'd be forgiven for thinking that. These photons are "discrete," not "discreet," in the sense that each of these energy packets has fixed attributes and measurements. Their traits are defined rather than on a spectrum, which is why these packets are known as quanta (because they're "quantized").

The energy in a photon can only be in certain multiples of a basic unit, which you can work out using Planck's constant (h) and the light's frequency (f) using the formula ***E=hf***.

Here, dear reader, lies the key difference between classical and quantum optics. In the former, energy can be assigned any value and exists in a continuous range rather than being discrete or quantized. Also, quantum optics honors the principles of quantum mechanics, which, as you know, do not jive well with classical physics. The essence of quantum optics is to help everyone understand the kooky, spooky stuff that goes on with light at the quantum level.

So here's a summary of the differences between both fields of optics. The classical view of optics holds that light occurs in continuous waves, while quantum optics sees light as individual particles called photons, with the ability to exhibit wave-like behavior, in line with the wave-particle duality of quantum physics.

Classical optics holds that any value can be taken on or assigned to light, whether that's in terms of momentum, energy, or other quantities, while quantum optics insists on discretion in light's values. One more thing quantum physics considers to be true of light is the entanglement theory. It posits that the light particles can and do become linked to each other so that they mirror each other regardless of how much space is between them.

It would be remiss to talk about quantum optics without touching on Quantum Electrodynamics or QED, the relativistic quantum field theory of electrodynamics. In case you feel that the previous sentence might as well have been summarized as "gobbledygook," here's a breakdown of what that means.

You already know quantum field theory is the study of the interactions between particles. Electrodynamics is about taking a closer look at *electrically charged* particles, in particular, and watching how they interact with light. As for the word "relativistic," it's from Einstein's theory of relativity, which, if you remember, states that no matter how fast you go, the laws of physics and the speed of light will remain constant for everyone and everything observing you.

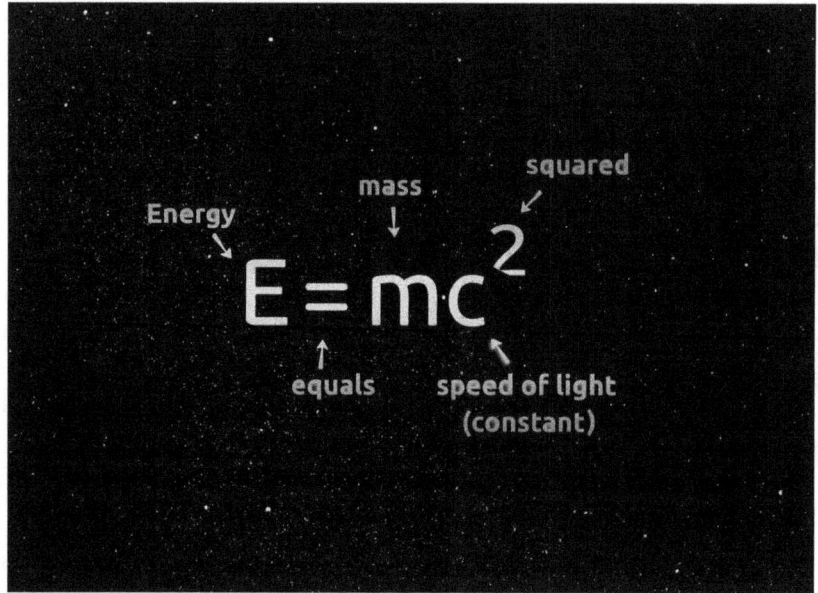

Einstein's theory of relativity states that no matter how fast you go, the laws of physics and the speed of light will remain constant for everyone and everything observing you.
https://www.needpix.com/photo/download/84526/einstein-formula-mathematics-equation-equations-formulas-free-pictures-free-photos-free-images

Now, putting all that together, QED is a combo of studying the smallest of things (quantum mechanics) and the fastest of things (special relativity) to illuminate how matter and light interact with one another.

Interesting side note: This theory is the very first of the lot in the quantum field, and it actually aligns well with Einstein's relativity theory. Using the language of mathematics, QED describes everything that happens with electrically charged particles as they exchange photons, and it's the quantum answer to classical electromagnetism.

According to quantum electrodynamics, the interaction between photons and matter (also called "coupling") concerns an energy exchange between both. It is a coherent process, meaning both matter and light are within the same frequency and phase, making it possible for the energy exchange to occur without any loss - and without it dying out or dissipating.

To make that simpler to understand, think of what "incoherent" interactions might imply - in this scenario, as energy is exchanged, it is lost as radiation or heat. Now, back to the matter of coherent energetic exchange between light and matter. In QED, particles on both sides have the same energy and momentum.

Now, there are four fundamental forces of nature.

1. **Gravity:** The force that pulls two objects together as long as they have energy or mass. Gravity also happens to be the weakest of the four forces, but it makes up for that weakness with the fact that there's not one thing in all the universe that isn't affected by it.

2. **Electromagnetism:** Then there's electromagnetism, which is what you find between electrically charged particles and create magnetic and electric fields. Unlike gravity, electromagnetism doesn't have as much range, and if you want to cancel it out, all you need are opposite charges, and it's over.

3. **Strong Nuclear Force:** Next, there's the strong nuclear force responsible for keeping a neutron and proton bound with the nucleus of an atom. It isn't called "strong" for nothing, as it's the most powerful of the four natural forces.

4. **Weak Nuclear Force:** Finally, there's the weak nuclear force which leads to nuclear fusion and radioactive decay and is only found with subatomic particles. According to the classical point of view, light and matter are seen as distinct things, and light is

only ever seen as a wave. However, according to QED, light and matter are seen as unified.

The Pioneers of QED

Who are the brilliant minds who pioneered this quantum field theory, and what did they contribute? Well, you already know about **Paul Dirac**, who was the first person to come up with the quantum theory of radiation and matter interaction. He also came up with the term "quantum electrodynamics," which he shared with the world in 1928. Dirac worked out the mathematical equation explaining what was going on with the movement of electrons and how they spin. He dubbed his explanation "the wave equation."

Next up was **Enrico Fermi.** Enrico came up with a brilliant formulation of Quantum Electrodynamics in 1932. How? He worked with the idea of virtual particles, using them to clarify how charged particles interact with light particles. Other honorable mentions are **Felix Bloch, Arnold Nordsieck, and Victor Weisskopf,** who would shed light on the problem physicists constantly came up against when it came to infinities in higher-order calculations.

Enrico Fermi
Argonne National Laboratory, , ATTRIBUTION-NONCOMMERCIAL-SHAREALIKE 2.0 GENERIC, CC BY-NC-SA 2.0 <https://creativecommons.org/licenses/by-nc-sa/2.0/>https://www.flickr.com/photos/argonne/50394459604

Yes, you want that broken down, don't you? You remember QED is about the interaction between light and matter at the quantum level, right? Well, physicists only predict these interactions through the lens of the perturbation theory, which you can think of as making a series of approximate calculations that become more accurate with each new iteration.

These three scientists found out that the perturbation theory was sorely lacking. You see, each time they wanted more accurate approximations, they would only wind up with answers that didn't add up. The answers were infinite, and that's why that phenomenon is called "the problem of infinities," which meant QED couldn't be trusted because it lacked consistency and, therefore, couldn't be relied upon.

So, Bloch and Nordsieck came together to figure out this problem and discover something to help them skirt around the infinities issue in a specific context. In this situation, when the charged particles emitted significantly low-energy light, the infinities would be canceled out, which meant that scientists' predictions could match their experiments. This solution is known as the Bloch-Nordsieck Theorem.

As for Weisskopf, he worked on his own and discovered there was another unique instance that allows you to avoid the problem of infinities: when the charged particles absorb light, that happens to be higher energy.

Was the problem of infinities ever sorted out? Yes. In the late 1940s, **Richard Feynman, Julian Schwinger, and Shin'ichiro Tomonaga** would independently come up with a solution, developing a version of QED that was dependable and accurate. Feynman offered his diagrams, Schwinger developed the action principle, and Tomonaga shared his ideas on renormalization. Feynman's diagrams are drawings showing how photons and electrons interact with one another by swapping photons, and they mathematically depict the odds of these particles interacting in a specific way. His diagrams work because he assigned some of them with a negative sign, which meant that after summing them all up, the infinities were canceled out, leaving behind answers that made sense.

Schwinger's action principle says that when you have a physical system, its action, which is a quantity physicists use to measure how said system changes over time, will always be the smallest or largest value possible. This rule helped him explain the movements and fields in

QED. That's not all Schwinger came up with, though. Using a method he dubbed "regularization," he found you could add a small number to the values to force the infinities to become finite, and then at the end of the calculation, he'd eliminate that same number.

Tomonaga's renormalization is another interesting method centered around the fact that when it comes to the numbers in Quantum Electrodynamics, such as the values for the electron's charge and mass, those figures are never fixed. Instead, they're in a state of flux, and their value depends on how small or how fast these traits are measured. So, Tomonaga would use his method to switch the numbers around to eliminate the infinities.

The Photoelectric Effect and the Compton Effect

This is a chapter about light, so it makes sense to talk about the photoelectric effect. What's that about? It's when the light leads to the ejection of electrons from a metal's surface, which was first noticed by Heinrich Hertz in 1887. This phenomenon was not explained until Albert Einstein offered an explanation in 1905.

Through the lens of classical physics, light is a wave that can have any amount of energy depending on how quickly it vibrates and how intensely or brightly it shines. The assumption was that the number of electrons ejected (and their energy) came down to these two factors, *but experiments proved otherwise*. What did experiments demonstrate?

1. The intensity of the light determined how many electrons would be emitted. In this case, more intense means more electrons.
2. The electrons' energy is a matter of the light's frequency. So, if the light has a shorter wavelength or higher frequency, the electrons emitted would have more energy.
3. There is a specific threshold below which no electrons can be ejected. This is the case when the light has a longer wavelength or lower frequency. Also, its brightness is not a factor at all.
4. Finally, the electron ejection process happens the moment the light connects with the metal. You know how you'd expect to hold a blade over the fire for a while before it finally gets hot? Well, that's not how it works when light hits metal because the emission is instant.

Einstein figured the results were what they were, thanks to the fact that light is a stream of photons, each with a set amount of energy determined by its frequency. The formula he came up with to explain a photon's energy is ***E = hf***, with E being energy, f being frequency, and h being Planck's constant, which has already been discussed.

Einstein also posited that each photon could only transfer energy to just one electron, and not only that, but the metal also needed a work function (a set amount of energy) to eject an electron. His formula for working out the kinetic energy contained in a single electron emitted by the metal is ***KE = hf - W***, with KE being kinetic energy, f being frequency, h being Planck's constant, and W being the work function.

The photoelectric effect shows the truth about light being able to act as a wave and a particle and the fact that light's energy must be quantized. It also demonstrates that light and matter do not interact with each other continuously and smoothly – but probabilistically.

Now that's out of the way, it's essential to focus on the Compton Effect, eponymously named after Arthur Compton. This effect was first observed in 1923. Arthur found that light can change its frequency (color) or wavelength and scatter off electrons, demonstrating that light truly acts as a particle.

Classical physics always assumed that scattered light would have the same frequency as incident light. To be clear, scattered light is the light that is produced as a result of bouncing off the electrons, while incident light is the light before the bounce happens. Classical physics was proven wrong once more, as experiments showed that:

1. The frequency of scattered light is much lower than that of the incident light. In other words, it has the lower frequency of the two lights.
2. The wavelength or frequency changes depending on the angle at which the scattering occurs. So, if the angle is larger, you can expect larger changes, and the smaller the angle, the smaller the changes.
3. These frequency changes have nothing to do with the light's intensity.

Arthur Compton confirmed the theory of light being a stream of photons. To understand the Compton Effect visually, scattered light will have a redder hue than the incident light. How does this play out? First, the photon strikes the electron, and in the process, the former gives some of its momentum and energy to the latter, causing the electrons to move faster and the photon to lose its own momentum and energy. So, you'll notice the photon losing its blue hue and looking more like red.

Arthur Compton confirmed the theory of light being a stream of photons.
https://picryl.com/media/arthur-compton-1927-91b473

Quantum States, Coherent States, and Squeezed States

Think of quantum states as mathematical explanations of the different results you get when you measure systems like photons, molecules, or atoms to track variables like spin, polarization, energy, etc. Quantum states are represented with symbols known as kets or ket vectors. You could write the quantum states of physical systems using combos of simpler states, also called *basis states*. These basis states have a set value for each property being measured.

A photon with zero energy is written as $|0>$. If it has 1 unit of energy, it's written as $|1>$. Another way to write out a photon's quantum state is by using a combo of two other kinds of basis states, where the photon is horizontally polarized ($|+>$) or vertically polarized ($|->$). There's also the photon being written with the basis states $\backslash L>$ and $|R>$, which stand for the photon either having a left circular polarization or a right one, respectively.

One thing to note about quantum states is you can't know all the traits of a physical system simultaneously because of the uncertainty principle.

Also, as a system's quantum state interacts with its environment or other systems, it changes with time. This change is worked out using the Schrödinger equation. The very process of measuring anything about a system is enough to change its quantum state, forcing it to collapse into or select one of the basis states. Which one? It's impossible to predict, at least for now.

So, when you're considering the quantum state of light, you'll learn about the probabilities you'll get from measuring its properties, and you won't be able to predict the wave-function collapse. Some quantum states are more useful or meaningful than others, like coherent states, squeezed states, and entangled states.

Coherent states are quantum states of light with special traits. They never fluctuate in time because their phase and amplitude remain constant. Their shape and size remain the same over time, and so do their direction and color. So, light in coherent states can't blend in with other colors. Also, coherent states are harmonious, meaning they will blend well with other photons or waves of light.

Look at lasers, for instance. (Laser is the acronym for "Light Amplification by Stimulated Emission of Radiation.") The stimulated emission is a quantum process. This is when a photon stimulates or excites a molecule or atom with the same energy to create a new photon with the same energy level, direction, phase, and frequency. The photons that are created are all the same and coherent. These coherent states make tracking the light's phase easier, which matters in measurement methods like metrology, spectroscopy, and interferometry.

Lasers are created when a photon stimulates or excites a molecule or atom with the same energy.
astroshots42Follow, ATTRIBUTION 2.0 GENERIC, CC BY 2.0
<https://creativecommons.org/licenses/by/2.0/>https://www.flickr.com/photos/astro-pics/8468331718

Entangled states can be generated by coherent states, too. Entangled states are quantum states where two or more physical systems are inexplicably correlated to each other, and these states are excellent for making advancements in the fields of quantum computation, communication, and information.

Squeeze states are quantum light states that have a lower uncertainty in one trait of light versus another to which it's specially related. These states can be created using mirrors, fibers, crystals, and other similar materials. The benefit of light in a squeezed state is that you can use it for far more precise measurements, better cryptography, faster computation, and more.

So, now that you understand everything about optics according to quantum physics, it's time to dive into the various quantum experiments that shaped the field of quantum physics. You'll get detailed explanations of the experimental setups, procedures, and observed outcomes.

Chapter 4: Quantum Observations, Experiments, and Their Interpretations

In this chapter, you'll learn more about some of the most famous quantum experiments that have led to the current state of quantum physics.

Young's Double-Slit Experiment (1801)

Young's Double-Slit Experiment demonstrated that light acts like a wave. Before Young pulled off this groundbreaking experiment, there were two tenets that scientists held onto strongly when it came to light: the corpuscular theory, which Isaac Newton postulated, and Christiaan Huygens's wave theory.

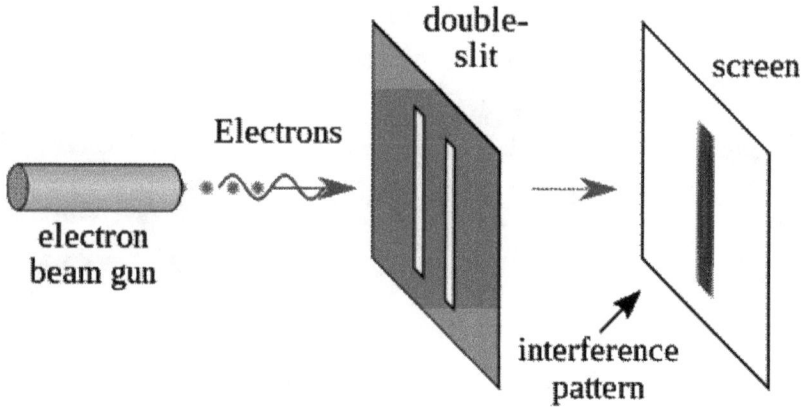

Young's Double Slit Experiment.
Original: NekoJaNekoJa Vector: Johannes Kalliauer, CC BY-SA 4.0
<https://creativecommons.org/licenses/by-sa/4.0>, via Wikimedia Commons.
https://commons.wikimedia.org/wiki/File:Double-slit.svg

Newton's theory of light was that it was made up of tiny particles, and they only ever moved in straight lines. Then, Huygens suggested that light was actually made up of waves and that these waves could bend, which meant they could affect and interfere with one another.

Back then, some scientists were more into Newton's idea than Huygens's. The way they saw it, Huygens's theory offered an obvious and far superior explanation for why light refracts and reflects than the wave theory. That didn't mean the corpuscular theory wasn't riddled with its own issues.

For instance, why can light diffract or bend around a slit or the edges of an object? Newton's theory could never explain that phenomenon. Why can light create colors when it comes in contact with thin films of oil slicks or soap bubbles? That was yet another question the corpuscular theory couldn't offer clarity on. It also didn't offer the best explanation for how light has the distinct trait of being unable to collide with each other whenever two or more beams are crossed with one another.

Young was taken with light and its nature. This physicist, also a physician, once saw light through the lens of the particle theory, but his investigations showed him the benefits of the wave theory, which did a good job of explaining the inexplicable. He found that light waves could actually interfere with each other.

"Interfere" in this context refers to the fact that light waves can cancel out and combine, depending on their relative positions. So, driven by his desire to learn all he could about light, he decided to do his famous Double-Slit Experiment.

Young kept it clean and classy. He didn't believe in complicating things for the sake of seeming impressive. The simplest way to replicate this experiment is to use an opaque object like a wall or a block, which has two slits carved in it. You would also need something to keep this object in place – and a monochromatic light (like a laser).

This light source should have something to support it, so the only reason it moves is *that you want it to*. The light should be directed toward the middle of the slits and positioned a half meter away from the double-slit object. On the other side of the object, there should be a smooth white wall or a screen a few meters away. When you've finished setting your experiment up, you'll notice dark and light bands showing up on your screen.

Laser light is excellent for this experiment because it can create a photon or more when it's powered with enough electricity, and those particles can emerge from the tiniest hole imaginable after a set period. Since the speed of light is not a variable but a fixed figure, it's possible to set a time for when the photons will appear on your screen.

If your laser's photons are created one after another, they'll appear as single light spots, proving they're particles. If they're waves, then it's natural to expect them to diverge or spread out as they push forward, and that means you'd expect to see a wide area of your screen being lit up – *but that's not what happens.* Photons being particles would imply that they should appear at two separate points on your screen or wall, but *that's not the case.*

Young didn't have access to lasers when he first carried out his experiment. He approached the process with the idea that light must be like water waves . . . and assumed that the light waves would travel from their source in the same way that ripples spread out when you drop a pebble in a lake. He also assumed that once the traveling wave hits the double slits, they'll become two distinct waves the moment they pass through the openings.

The light wouldn't show up as waves on the screen when Young did this experiment. Instead, it became obvious that the photons were hitting the screen on their own. Also, one of these particles could interfere with itself in the same way a wave would, according to classical physics. The photon could split once it got to the double-slit, only to reunite its parts once it hits the screen.

The Photoelectric Effect (1887)

Picture it: A dark lab in Germany. The year is 1887. Heinrich Hertz, a 30-year-old, is hard at work, observing what happens when he shines an ultraviolet light beam onto a plate made of metal. He watches, fascinated, as the metal plate shoots off sparks. But it isn't really the emission that has his attention. You see, it's well known that metals are excellent electricity conductors since the electrons in this material aren't so rigidly connected to the atoms – meaning it won't take much to dislodge them with the right amount and intensity of energy.

So Hertz had a puzzle before him. He realized that the frequencies of the light bursts that made electron emission possible depended on the metal in question. He also noted that when he would turn up the light's brightness, there were more electrons emitted. Still, there wasn't a correlated increase in energy. When he used higher frequencies of light, he'd get electrons with increased energy. Still, there wasn't a commensurate increase in the number of electrons produced.

This phenomenon would eventually be dubbed the photoelectric effect, one which a young Albert Einstein would be able to fully explain later on in 1905. The photoelectric effect used to be quite a conundrum for classical physics, but it would also be one of the first wins Einstein scored during his career. This effect proves the fact that light is quantized.

Here's a simpler breakdown of the effect. When you shine a light on metal, electrons are emitted and then absorb the light. When these particles have enough energy, they'll set themselves free from the metal. Classical physics assumed light was only a wave and that there's no specific amount of energy that it swaps with the metal. The classical assumption, therefore, is that when you shine the light on the metallic object, the object's electrons absorb the light, and the energy gradually increases until there's enough to cause the electron emission process. Also, it was expected that when you shine even more light on the metal, you should notice the emitted particles move with a much higher kinetic energy. On the flip side, if the light is too weak, there's no way the metal can throw off electrons unless enough time passes.

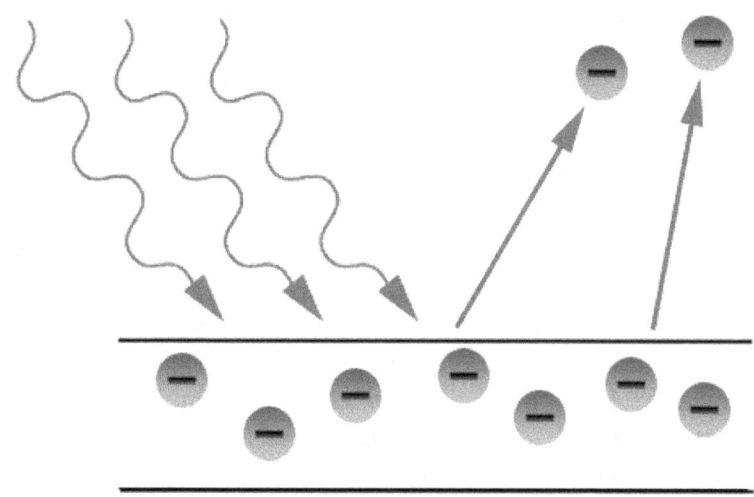

The Photoelectric Effect.
ATTRIBUTION-SHAREALIKE 3.0 UNPORTED, CC BY-SA 3.0
<https://creativecommons.org/licenses/by-sa/3.0/deed.en>https://upload.wikimedia.org/wikipedia/commons/7/77/Photoelectric_effect.png

Here's the thing: the experiment showed that these assumptions were false. The particles emitted the moment the light hit the metal, and regardless of how intensely bright or dim the light was, there would still be an immediate emission of electrons. So, the only thing you need to instigate that process is the frequency of the light rather than the intensity. Einstein would find an apt explanation inspired by Planck, and that's why he claimed light was quantized into photons, which meant it acted like waves and particles.

So, what's really happening when light strikes the metal object? The photons (light particles) collide with the loose electrons, and each electron swallows up each photon. When the photon has more energy than the object's work function, the electron is emitted. The formula that demonstrates this is written like this: $h\upsilon = W + K$, with W being the work function of the metal and K being the emitted electron's kinetic energy. To be clear, "work function" refers to the least amount of energy you would need to knock an electron free from a material. The material is usually metal.

The Stern-Gerlach Experiment (1922)

What's the Stern-Gerlach experiment about? Well, it was the experiment that showed scientists that spin is a real thing – and no, not "spin" in the public-relations-Edward-Bernays type of way! To put a rather pedestrian spin on the definition of this concept, think of a spinning top. That's a rough analogy of how the tiny little particles that make up life move.

Otto Stern and Walther Gerlach were responsible for the revelation of this angular momentum. How did they discover it? Well, they had a beam of silver atoms channeled through magnetic poles, allowing it to hit a screen. What's interesting about this is the fact that silver has 47 electrons, but only 46 of them are arranged in a symmetrical cloud, which means they're not responsible for the atom's spin. What about the 47th electron? It's either in its 5s state or 5p state.

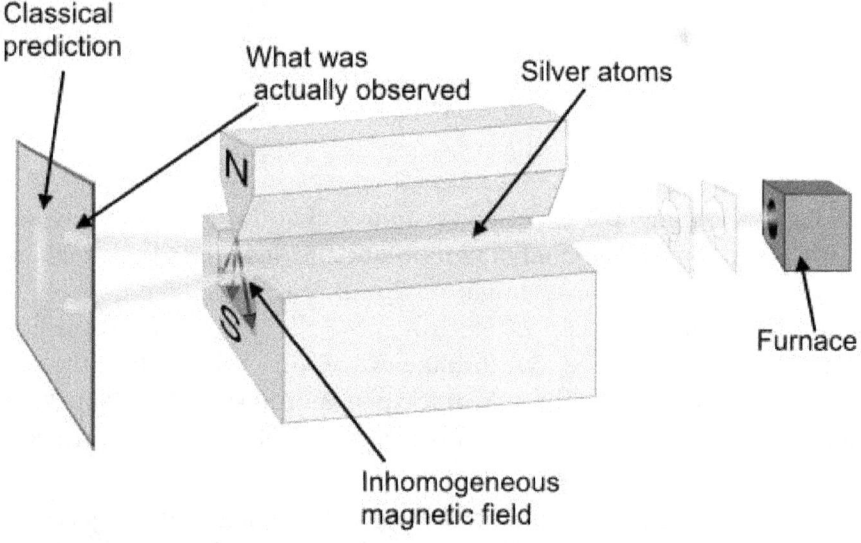

Theresa Knott
Theresa Knott from en.wikipedia, CC BY-SA 3.0 <http://creativecommons.org/licenses/by-sa/3.0/>, via Wikimedia Commons https://commons.wikimedia.org/wiki/File:Stern-Gerlach_experiment.PNG

An atom is not unlike a solar system but on a microcosmic level. In the middle of the atom is the nucleus, acting like the sun in the sky. As planets surround the sun, so do electrons surround the nucleus. Now, the electrons must travel through or occupy certain energy levels or lanes, which you can call "states." So, when silver's 47th electron is in the 5s state, there's no wobble or tilt as it travels a straight path while orbiting the nucleus, which means its angular momentum or spin has a value of 0.

Now, when this special electron is in the 5p state, its path is tilted as it travels around the nucleus, which means its angular momentum has a value of 1. The 5p state means the electron moves in one of three directions. If it's tilting down, the value of the spin is -1. When it's titling up, it's +1. When there's no tilt, it's at 0. Are you confused about why the electron with no tilt is still in its 5p state rather than 5s? The 5p electron with no tilt may seem the same as the 5s, but it's actually on a different path.

Stern and Gerlach had thought they'd find one or three spots on the screen as they beamed their silver atoms through the magnet poles, but there were only two. Scientists would find this puzzling for a whopping three years, as they tried to find a theory to explain what was going on. The answer was discovered in 1925 by George E. Uhlenbeck and Samuel A. Goudsmit, whoh postulated that the electrons had intrinsic angular momentum.

Besides the angular momentum of the electrons, as they spin around the nucleus, they found there was an inner angular momentum or spin. The Stern-Gerlach experiment demonstrated that the beam of silver atoms splits in two, and this split depends on the way the 47th electron spins. Scientists learned that there are two kinds of spins, one up and the other down.

There's nothing in classical physics that talks about the idea of spin. It's only a quantum mechanical phenomenon. Even the analogy of the Earth spinning on its axis while spinning around the sun isn't the best one to explain spin. Also, if you could somehow stop an electron and put it in a state of inertia, it would still have its intrinsic spin. You can't take that away from it.

The EPR Paradox (1935)

Also called the Einstein-Podolsky-Rosen Paradox, this is an interesting thought experiment that's meant to illustrate an intrinsic paradox that scientists tried to wrap their minds around when quantum theory was yet in its infancy. It's one of the best demonstrations of the concept of quantum entanglement. So what's it about?

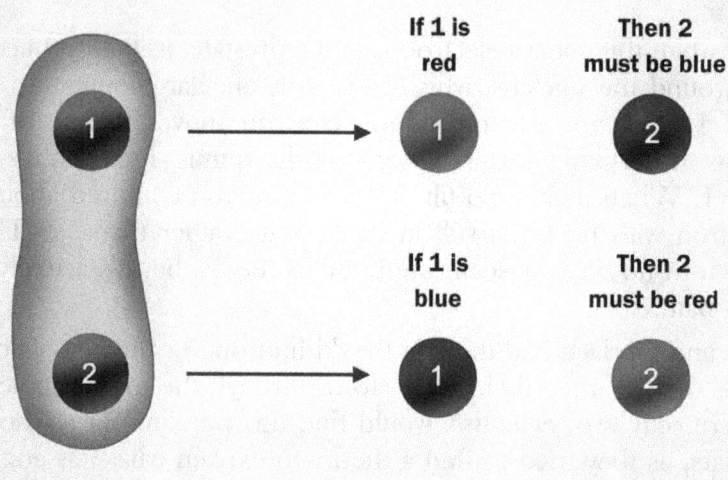

Measuring a pair of entangled photons.

Picture two particles entangled with each other. Until you measure each of them, they remain in a state of uncertainty. When you measure one, then it takes on a state of certainty, and so does the other yet unmeasured particle it's entangled with. As you've learned, this magic is possible because they both communicate with each other at speeds beyond the speed of light, which immediately calls into question Einstein's Relativity Theory.

This EPR Paradox was something Albert Einstein and Niels Bohr both exchanged intellectual blows over. You see, Einstein didn't want to accept quantum mechanics with open arms, while Bohr and his supporters were developing this field even further. What's interesting about this is Bohr's work was actually based on something Einstein had begun.

EPR Paradox & Bell Inequality
Einstein, Podolsky, Rosen experiment (Continued)

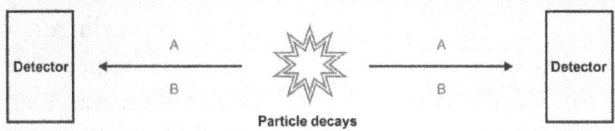

- EPR said each particle is "real" -- it is "A" or "B" no matter what any detector says.

- Quantum Mechanics predicts two particles are "intertwined" in one wave function.
  Neither particle is in a definite state (A or B) until it is detected.

- John Bell (1964) constructed a simple inequality which could be measured to decide who is right.

- Result: Quantum Mechanics is right.

The EPR Paradox.

Einstein teamed up with Boris Podolsky and Nathan Rosen, and together, they created the EPR Paradox with the intention of demonstrating the inconsistency of quantum physics with the laws of physics as they were known at the time. They didn't have the means to put their thought experiment into practice back then.

A few more years passed, and David Bohm would change things up with the EPR paradox example, with a view to making it easier to understand. Even the top physicists' unstable spin 0 (zero) decays of the time couldn't quite explain the paradox. In Bohm's version, a particle with an unstable spin 0 decays (or transforms) into two other particles, different from one another and moving in opposing directions, one clockwise and the other counterclockwise.

Since the original particle's spin was 0, the new particles have the same value in their spins. If one of them has spin +1/2, the other will have spin -1/2. Also, the Copenhagen interpretation of quantum mechanics holds that these particles don't have definite states until they're measured, as there's an equal chance they'll either have a negative or positive spin.

Schrödinger's Cat (1935)

Now it's time to get to know Schrödinger's cat a little bit better. This was a thought experiment that scientists used to test out other quantum ideas, and it was born in the brilliant mind of Erwin Schrödinger, in 1935. He came up with this because of how quantum mechanics were explained according to the Copenhagen interpretation, which was that particles in the context of quantum mechanics are in existence in every state you can imagine simultaneously, *unless and until they are observed*, and only then do they select one out of the plethora of states to adhere to.

For instance, you could have a light bulb that, when lit, could either be red or green. When you're not looking at the light bulb, the Copenhagen interpretation of quantum mechanics would have you assume that the light the bulb emits is both colors, red and green. Yet, when you do look at the light, it will have to be one or the other, not both. This wasn't something Schrödinger agreed with, and this was why he'd introduce the world to his cat experiment.

Here's the thought experiment in a nutshell. Pretend, for a moment, that you have a cat. Not only that, but you also have a small piece of some radioactive substance, which is something that's unstable and emits particles randomly. Now, you place both your cat and this radioactive object into a box and seal it in.

Also, you rig the box with a device that will release poison into it. This is no ordinary poison. It can only kill the cat if the device picks up on one of the particles emitted by the radioactive substance. Once the radioactive substance decays, it emits particles, triggering the device (a Geiger counter) to go off. Once triggered, the device releases poison that leads to the totally unjust and horrendous death of your cat.

When you consider the observation theory and bring it to bear in this experiment, since no one's watching the cat (remember it's sealed in a box, and you don't have x-ray vision), the cat has to be both dead and alive. Why? The radioactive substance will decay and won't decay. The poison will and won't be released. At least, not until you open up the box to check on your cat, at which point it will then be one or the other. Looking in on your cat is the same as "measuring" the outcome, which is the process thought to force the cat to either be okay or give up one of its nine lives. Schrödinger's point was that this was an absurd thought and an impossibility in real life for the cat to be in both states. He demonstrated with this thought experiment that the cause of wavefunction collapse has nothing to do with whether there's an observer or not.

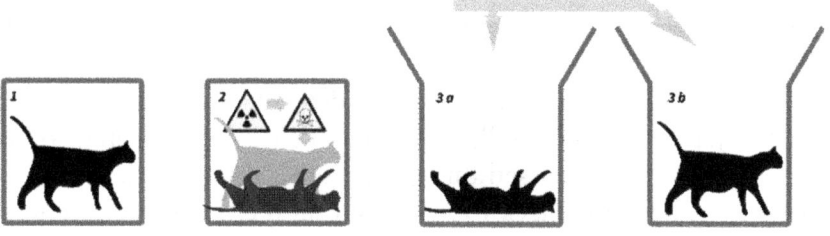

Schrödinger's cat experiment.
Master of the Universe 322, CC BY-SA 4.0 <https://creativecommons.org/licenses/by-sa/4.0>, via Wikimedia Commons https://upload.wikimedia.org/wikipedia/commons/7/75/Physics-3864568_960_720.png

As Nobel Prize winner and physicist Robert Penrose would later write in his book *The Road to Reality*, the cat being both dead and alive simultaneously is absurd when looked at in the context of the physical world. He pointed out there's a 50-50 chance of the cat being dead or alive, strictly physically speaking, and that this is proof of the fault in all interpretations of Schrödinger's cat that seek to prove the cat is in both states at the same time. Schrödinger demonstrated how impossible it is for things to exist in a state of superposition in real life. Unless, of course, there's more to life than meets the eye.

The Delayed Choice Quantum Eraser (1998)

Early in 1998, Yoon-Ho Kim, R. Yu, S. P. Kulik, Y. H. Shih, and Marlan O. Scully worked on the Delayed Choice Quantum Eraser experiment. The point of this experiment was to probe further into the results of the double-slit experiment, as well as where quantum entanglement ultimately leads.

The scientists worked with a Spontaneous Parametric Down-Conversion laser beam and a Beta-Barium Borate crystal (also called a BBO crystal). The laser beam they chose was a powerful one, which, when directed onto the right crystal, will cause the light to split into pairs of much weaker photons than contained in the original beam.

The photons being shot out of the SPDC laser beam and onto the BBO crystal are in pairs. They are entangled, so whatever you observe by studying one of the pair of particles is definitely happening to its twin, regardless of their distance from each other. When the photons are shot at the double-slit wall, each photon in a pair could choose to pass through one slit or the other.

Beyond the slits, there's a device that detects which slit every particle or photon passes through. Still, the catch is you can turn it on or off only after the photons have passed through these slits – and this is where the fun begins. When you check to see which slit the photons chose to pass through, you'll notice they have particle-like traits because they'll either pass through one slit or the other, but never will they go through both.

What about when you don't track the paths of the particles with the device? In that case, the photons become like waves, which means they pass through both slits simultaneously and create a rippling pattern. The even weirder thing is whatever you decide seems to determine how the photons *acted in the past.* You might as well call these photons psychic because it's like they knew if you'd use the device to track their path or not. This is why it's called "delayed choice." The "eraser" bit suggests that one outcome or nature of the photons is erased in favor of the other.

What's really going on with these psychic particles? Well, they're not exactly psychic, nor do they have the ability to go back in time and change things. What this experiment does is present a challenge to the way everyone has always assumed time works. The classical idea of time is linear. In other words, your present is supposed to be the sum total of your past, and your present determines your future.

This experiment suggests time may not be linear and that all three faces of time are intertwined with one another in ways that continue to baffle scientists even now. As fascinating as all this is, some insist that the retrocausality suggested by the experiment is misunderstood.

Interpretations of Quantum Mechanics

There are various ways to interpret quantum mechanics and its theories. Here's a quick look at some of them.

The Copenhagen Interpretation: Of the many interpretations out there, this is the most accepted one, founded on the idea that particles act in line with the probability wave notion – and that superposition is valid. According to this interpretation, the act of observation of measurement forces wave function to select or collapse to one state only (probability wave), and particles can be in more than one place at the same time (superposition).

The Many-Worlds Interpretation: The idea behind this interpretation is that there isn't a fixed history or future and that multiple versions exist because there's more than one universe or world. So, in the quantum world, the universe splits into several more with each event that occurs.

The Pilot-Wave Theory: The distinguishing feature of this interpretation is that there are hidden variables in the quantum world, and this is why all the random, unpredictable, spooky action in quantum mechanics happens. This is also called the *De Broglie-Bohm Theory*.

Quantum Bayesianism: Also known as *QBism*, the interpretation suggests your beliefs about a system's state are what play out as the wave function.

Objective Collapse Theories: These interpretations have the premise that the wave function collapse isn't relegated to the quantum world but is physical and real.

Relational Quantum Mechanics: Through this lens of interpretation, you assume that the same series of events may be observed and interpreted differently depending on the context.

Transactional Interpretation: The wave nature of the particles in the quantum world matters when viewing quantum affairs through this context, and waves and particles are equally important as they complement each other.

What's the point of all these interpretations? They're all attempts by philosophers and physicians to describe the *true nature of reality*. Where one interpretation falls short, another may pick up the slack and offer explanations that make sense.

Chapter 5: Quantum Reality and Consciousness

You can't learn about quantum physics without beginning to question the nature of reality. Quantum reality and consciousness remain matters of intense debate, drawing scientific and non-scientific minds alike. In this chapter, you'll open your mind to the thought that consciousness has a deeper role in life as you know it than you could ever imagine.

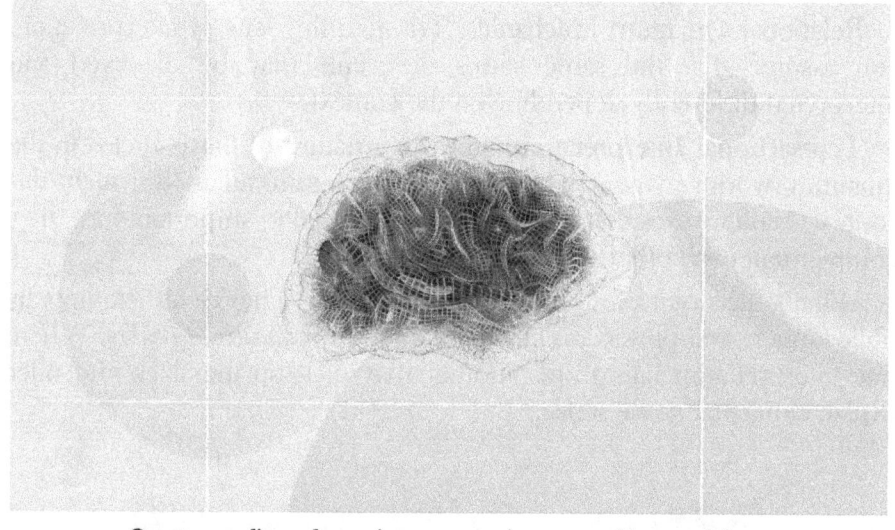

Quantum reality and consciousness remain matters of intense debate.
https://www.pexels.com/photo/an-artist-s-illustration-of-artificial-intelligence-ai-this-image-represents-how-machine-learning-is-inspired-by-neuroscience-and-the-human-brain-it-was-created-by-novoto-studio-as-par-17483868/

Quantum Mind Theories

Quantum mind theories attempt to explain consciousness as clearly as possible so that humanity understands itself better. Think of how your mind works. Do you assume it's all just neurons doing their thing in your brain? Well, quantum mind theories suggest there's more at play. The inner workings of your mind depend on quantum rules, and this is what makes human consciousness so dynamic, rich, and fascinating.

Quantum mind, also called quantum consciousness, is a set of theories or hypotheses that put forth the idea that superposition, entanglement, and other quantum physics events are what create consciousness. Consciousness, a subjective and personal thing, is a tough nut to crack using quantum physics, but there have been a few interesting quantum mind theories put forth that seem to explain it.

David Chalmers is the philosopher who coined the term "the hard problem of consciousness." What is this problem? Well, how do the physical actions of your brain cells cause your subjective experience of life, if they even do? Why do you feel something as one thing rather than as another? What's responsible for your inner experience of life, which is different from others' inner experiences? Why do you even have an inner life in the first place?

David Chalmers coined the term "The hard problem of consciousness."
TEDxSydney, ATTRIBUTION-NONCOMMERCIAL-NODERIVS 2.0 GENERIC, CC BY-NC-ND 2.0 < https://creativecommons.org/licenses/by-nc-nd/2.0/>
https://www.flickr.com/photos/tedxsydney/5779378540

The "hard problem of consciousness" is working out how and why living beings have subjective, conscious experiences, also called "*qualia.*" It's one thing to know how and why certain aspects of the human brain make it possible to tell things apart from each other, to process and understand information, and to carry out specific actions. The explanations for those things are rooted in functionality and behaviorism, but the same doesn't apply to the hard problem of consciousness. Here's a look at some of the quantum mind theories that have been put forth to solve this hard problem.

Bohm's Implicate Order

You know matter is whatever has weight and occupies space. As for consciousness, it's the ability you have to be aware of yourself, others, and the world around you. It's being able to feel, perceive, and think. *Bohm's Implicate Order* is an attempt to find the thread that ties consciousness, matter, and quantum physics.

According to Bohm, there's more to reality than meets the eye. There's a deeper level where everything is connected to everything else, thanks to quantum phenomena. Bohm referred to this deep level as the Implicate Order – to reflect the fact that this level is hidden from regular observation. Being a creative person, he had another term for the Implicate Order: the *Holomovement*, which describes a movement that's complete or whole.

To conceptualize the Implicate order, think of it as an ocean that stretches as far as the eyes can see. You notice that this ocean is full of waves, and each of those waves represents a possibility in the quantum field. Bohm believed that the waves could overlap with each other, and this interference creates intricate patterns that you pick up on as consciousness and matter in the reality he called the Explicate Order (the visible, individual ocean waves with no overlap).

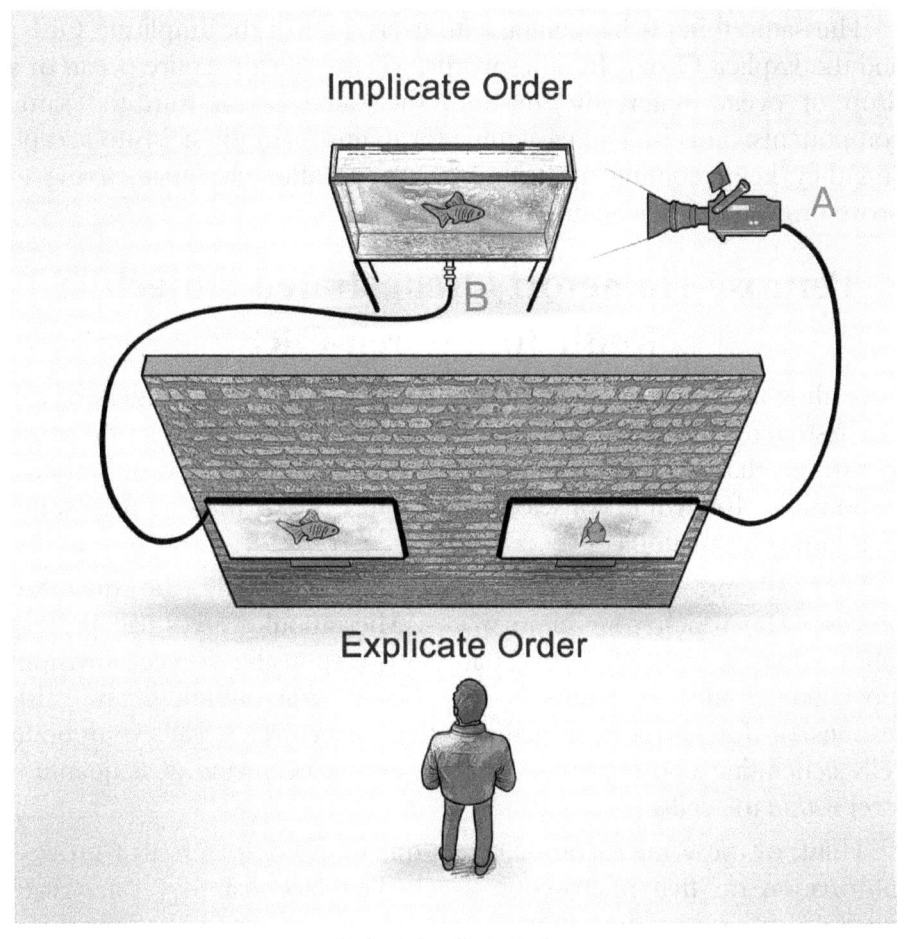

Bohm's Implicate Order.

Bohm says that matter and consciousness aren't to be seen as separate phenomena but as sharing the same foundational reality while presenting differing aspects. In other words, they're both born from the Implicate Order and reflect it. Consciousness, therefore, is the Implicate Order reflecting itself to itself, while matter is the manifestation of the Implicate Order in space and time.

The Implicate Order is ever in motion and always creative. It's in the business of crafting new forms and making room for new possibilities. The analogy Bohm drew was that of a hologram, which, if you haven't seen one in the movies, is a 3D image generated by lasers. When you break a hologram up into bits, you'll discover perfectly whole versions of the original hologram in each piece.

The same thing is happening with every part of the Implicate Order and the Explicit Order. In other words, whether it's the entire ocean or a drop of ocean water, they're both the same ocean with the same components. This isn't an easy thing for mainstream physicists to accept, and they keep coming up against challenge after challenge, trying to prove or even test this quantum mind theory.

Penrose-Hameroff Orchestrated Objective Reduction (Orch-OR)

According to Roger Penrose and Stuart Hameroff, the founders of the Orchestrated Objective Reduction (Orch-OR) theory, it makes no sense to assume that the neural networks in your brain are the only things responsible for your consciousness. They hold that there's some quantum computation going on as well.

Penrose and Hameroff say microtubules contain the quantum processes on which your brain works. Microtubules are small protein tubes within your neurons or brain cells, responsible for cell division, movement, and communication. These microtubules are also responsible for giving your neurons their structure. It isn't your brain cells generating consciousness but the events occurring at a quantum level *within* the cells.

Think of these microcomputers as quantum computers that process information on that microscopic level. The fact that they can create quantum superpositions implies these "computers" can be in a variety of states at the same time until the observer effect kicks in. They're so amazing they also create quantum entanglements, allowing them to connect with one another and create changes across space.

The thing about these states is they switch up their functions depending on the environment. They're subject to decoherence, where they collapse into one state. This is the very reason quantum computers have to run in environments with low temperatures and far from all disturbances.

Penrose and Hameroff claim your brain can avoid the decoherence effect, holding on to the coherence within its microtubules for an impressive period. With your memory, senses, and bureau structures, you can dictate the processes going on in your microtubules at the quantum level. But the question is, what connection do the quantum neural processes have with consciousness? How do they create it? Enter objective reduction (OR).

OR is a version of quantum collapse thanks to spacetime's nature rather than the observer effect or decoherence. What's space-time? It's the tapestry of the universe, the combination of space and time, that creates a fourth-dimension spectrum or continuum.

Once the microtubules are super-positioned, and things hit a level of instability, that's OR. Superposition is forced to collapse into one state, and this process is the birth of consciousness. The collapse, fortunately, doesn't need any observer to occur, as it happens on its own. Also, once it happens, you can't undo it or reverse it. Would you like to try to compute which choice it will make? You can't. There's not one algorithm that exists that could predict what's going on, and you could say this is the explanation of concepts such as creativity and free will.

Why did Penrose and Hameroff refer to this as orchestrated objective reduction? The way they saw it, your brain is what determines the location in space and the point in time of these OR occurrences in the microtubules, which leads to conscious moment after conscious moment, or what you'd call a "stream of consciousness." These scientists also held that the Platonic values are rooted in the framework of spacetime, which includes ethical values, mathematical truth, and aesthetic beauty.

The Quantum Zeno Effect

Another name for this effect is the Turing Paradox. This effect is about the fact that particles and other quantum systems could be forced to collapse into a specific state or "frozen" by measuring it as frequently as required, which keeps it from being superpositioned.

Remember, at the quantum level of existence, superposition is the order of the day because all particles are in all states at the same time until there's a wave-function collapse forcing it into a specific, unchanging state. According to the Quantum Zeno Effect, when you keep your eye on a particle long enough, you force it to remain in its original state. It loses its ability to change.

Here is a simplification of the effect in action. You're on YouTube. In your mind, the white line that indicates how much of the video has loaded is in a race against the red line, which shows how much of the video you've watched so far.

Unfortunately, the red line has caught up with the white one, and now, thanks to your internet working at the neck-breaking speed of a snail, you're forced to wait for the video to load so you can resume watching. You get sick and tired of waiting for the white line to load some more already. You're eyeing it like a hawk.

According to the Quantum Zeno Effect, the fact that you keep checking on the video's loading progress is the real reason it's not loading. It's like with the proverbial watched kettle that never boils; the only thing boiling is your impatience. Thankfully, your internet service provider can't use this as an excuse for why your videos get stuck buffering at the good bits.

What's with "Zeno" in the name of this effect? Have you ever heard of Zeno's arrow paradox? This is a puzzle of sorts, dating all the way back to ancient Greece. According to Zeno, the Greek philosopher of Elea, when you look at an arrow in flight at any point in time, it appears not to be moving. His argument, therefore, was that the flying arrow isn't actually moving.

Time is a series of moments or instances, and there's no motion in each moment or motion. Therefore, the arrow is still. It's the same thing with the Quantum Zeno Effect, as it appears that quantum systems operate the same way by freezing upon constant measurement or observation.

Now, what connection does the Quantum Zeno Effect have with consciousness? The idea is that all consciousness is affected by this effect, and by consciously observing these processes, you can influence them, freezing them into a state and thereby keeping them from changing.

For instance, if your consciousness is the result of the quantum superpositions in your brain, and you were to somehow monitor these superpositions consciously, that would keep the super-positioned processes from changing, which could be a viable explanation for how physical processes generate consciousness in the first place.

Consciousness, Spirituality, and Psychology

Being conscious is being aware. It's knowing you exist in space and time. Consciousness is the stuff that makes you feel like a real person, alive and kicking, different from the people and other creatures around you. Yet, as clear and obvious as it seems, certain aspects of consciousness can be rather difficult to pin down. It isn't yet fully understood how the brain and its neural processes can create consciousness, nor how consciousness is tied to your perception of the physical world. Some suggest that consciousness is not a product of the brain but that the brain itself and everything else in the observable world is the creation of consciousness.

People like the Dalai Lama see a connection between consciousness, spirituality, and quantum physics. The way he sees it, every atom in your body is an inextricable part of everything that makes up the world. You're literally made of star stuff. Your body has carbon, nitrogen, and oxygen, elements that were forged in fiery stars over 4.5 billion years ago. You're also intrinsically connected to every other thing on Earth, as you're made of energy like everything else on Earth.

People like the Dalai Lama see a connection between consciousness, spirituality, and quantum physics.
Yancho Sabev, CC BY-SA 3.0 <https://creativecommons.org/licenses/by-sa/3.0>, via Wikimedia Commons
https://upload.wikimedia.org/wikipedia/commons/2/2a/The_14th_Dalai_Lama_FEP.jpg

You know how difficult it is for spiritualists and scientists to see eye to eye. If you could travel back in time through some awesome quantum process to the Middle Ages and even the Renaissance, you'd witness this war between head and heart in real-time. Back then, any scientific progress was deemed dangerous, demonized, and was enough cause for murder.

Over time, the pendulum has swung to the other extreme, with spirituality being mocked by the world of science. So, how fascinating is it that, finally, there may be one thing where experts in both fields see eye to eye, especially when it comes to quantum physics and Buddhist philosophy?

Quantum physics demonstrates there's a world beyond the physical, one made of energy. Buddhists also agree with this, as their religion makes it clear that the physical must be transcended to give your full attention to your consciousness, which is what gives life its form and meaning in the first place.

This lends further credence to the quote by the late, great 17th-century French philosopher René Descartes: *"Cogito, ergo sum,"* meaning, "I think, therefore I am." It's your thoughts and your consciousness that shape your experience of life. This is also the foundation of many psychological practices, which seek to alter the mind's assumptions about life to help patients live as the versions of themselves they'd prefer to be.

The idea of consciousness being the true sculptor of life isn't something only Buddhists know. For instance, Amit Goswami of the University of Oregon also backs up the idea that micro-particles will change the way they act *depending on your actions as the observer*. This is a point that has already been clearly explained in this book.

Scientists and spiritualists have put down their blowtorches and pitchforks to agree, for once, that you and the world around you are defined by your thoughts and emotions.

This proposition is a challenging one for those whose minds aren't flexible and who prefer to follow orthodox ways. The implication of all this could be summed up by this wonderful quote by R. C. Henry in *The Mental Universe*: *"If we think about the possible connection between quantum physics and spirituality, we can see that the mind would no longer be that accidental intruder in the realm of matter, but would rather rise as a creator and governing entity of the realm of matter."*

Observation

Each time you interact with the quantum system, you are observing or measuring it. By using a macroscopic device such as a detector to look at an atom or a photon, you cause the wave function collapse that moves the atom or other particle from a state where it's everywhere and everything at once into a single form and location.

The act of observing the particle moves it from the realm of the indefinite to the definite. While you may understand this theoretically, it remains one of the most perplexing things that quantum physicists struggle to grasp.

It is one thing to understand the observer effect, but it's another thing entirely to know why it happens. What is it about someone being consciously aware of an atom that forces it to crystallize into a specific state? What even counts as an observer? Does it have to be someone with consciousness, like a human being, or could it be a device monitoring the particles on its own with no interference?

If you look at this phenomenon through the lens of the Copenhagen interpretation, the von Neumann-Wigner interpretation, as well as the many-minds interpretation, they all agree on one thing: An observer's consciousness is the key to forcing a wave function collapse to happen. You would be forgiven for assuming they're basically saying you have superpowers.

Other interpretations of this phenomenon do not suggest that the observer's consciousness has any relevance in causing the collapse of the wave function. According to these theories, collapse is something that happens objectively. When that's not the case, then it must be an illusion of sorts, which is a result of the particles interacting with their environment.

If you choose to view the collapse of wave function through this particular school of thought, then you have to stay away from the problems and paradoxes that come up as a result of introducing the element of consciousness. In other words, you would have no business with Schrodinger's cat.

It's not a new thing to suggest that consciousness plays a huge part in the crafting of physical reality. In addition to Buddhism, Taoism and Hinduism also have their own takes on this process. They suggest that the world as you observe it is an illusion, also known as *Maya*. According to these religions, there lies a "true" and "real" reality, so to speak, beneath the physical world, which is consciousness itself.

Those who follow these spiritual paths refer to this consciousness as Brahman, the Buddha nature, or the Tao. Reading their religious texts, it becomes clear that intention and thought are the only ways to influence the physical world. So, when you combine that idea with the observer effect, it becomes clear that whatever you observe is a result of your thoughts, emotions, and expectations.

Moving away from traditional Eastern philosophies and onto esotericism, mysticism, and occultism, the claim is that the physical world is a creation of the spiritual one. These forms of spirituality also agree that the way to alter your physical life is by using the power of intention and thought.

Other modalities to achieve this influence include meditation, prayer, visualization, magic, rituals, etc. These practices are meant to assist you in harnessing consciousness and molding it to achieve whatever goals you desire. This is also the logic behind how the impossible is accomplished, such as healing terminal diseases, receiving divine protection and timely provision, or inexplicable transformations.

Consciousness and the Quantum Field

You've become best buds with Schrödinger's cat. It's time to meet someone new: Wigner's friend.

Who's that?

Well, it's more like, *what's* that? It is a thought experiment that's a twist on Schrödinger's cat. Before you get acquainted, you should know that consciousness isn't an individual thing.

Every thought and sensation you've ever had or will ever have, every image you've conjured or come across, and every feeling you've experienced all spring from consciousness. These things also return to consciousness in the same way subatomic particles behave when it comes to the quantum field.

Now, back to the cat. It's in a sealed box, and its life depends on whether or not a radioactive atom expels poison that kills it. Remember, this cat is alive *and* dead as long as the box stays sealed. Wigner's friend comes over to observe this experiment. They have no idea whether the cat is alive or dead. When Wigner's friend opens the box and looks at the cat, that very act forces the wave function to collapse, and this means the cat is either dead or, better yet, alive.

Wigner's friend forces one and all to question what role consciousness may have had in the cat's fate, if any. In other words, is it possible that your mind is so powerful it affects events on a quantum level? This thought experiment clearly demonstrates the interplay between consciousness and the quantum world. You could say your new friend validates the observer effect, double-slit experiment, delayed-choice quantum eraser, and quantum Zeno effect.

The connection between the quantum field and consciousness is something that physicists continue to explore. For instance, Dr. Dirk K. F. Meiher, a professor at the University of Groningen in the Netherlands, thinks of consciousness as being within a field that's found around your brain, residing in a different dimension. He proposes that your brain pulls information from this field as needed through the quantum mechanism of entanglement and other quantum activities.

Not only that, but Meiher also believes the field is no different than a black hole in certain ways and is able to draw information from dark energy, Earth's magnetic field, and other interesting sources. You could call this field a meta-cognitive domain or a global memory space, very much like how the digital cloud saves all sorts of information on your behalf, ready for you to retrieve what you need when you need it.

The implications of Meiher's suggestions are significant because if consciousness isn't something your brain generates, and if there's a larger field from which it comes, that could only mean humanity must begin to question the material way it views the world.

You have to wonder about whether or not you have free will, what this implies about your identity, and what is really real. That said, you should now see the importance of self-awareness. It doesn't hurt to know what you're putting out with your thoughts and emotions since these will automatically draw corresponding experiences and other effects from the quantum field into your life, for better or worse.

Self-Awareness Exercises

If you want to become more self-aware, then look no further than practicing mindfulness meditation. Sure, it's a great tool to help you bust stress and stay healthy, but it does a whole lot more than that for you. You will experience a profound connection with the present, moving from moment to moment with full awareness rather than remaining stuck in past regrets or future worries. Perform the following exercises each day for phenomenal results.

Body Scan

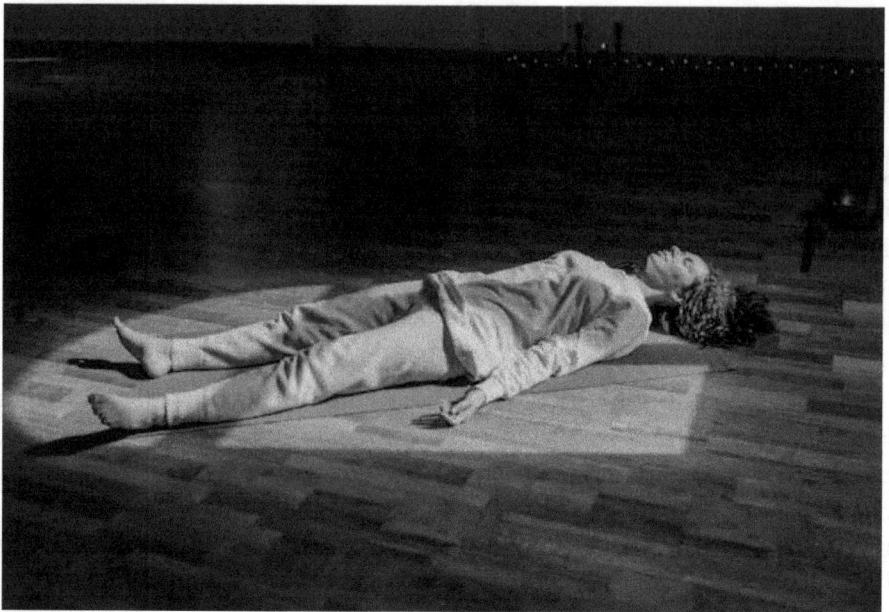

The body scan meditation is a great way to tap into your self-awareness.
https://www.pexels.com/photo/woman-lying-down-on-floor-relaxing-and-meditating-6998214/

1. Lie or sit comfortably. This should be a position you can hold for at least ten to fifteen minutes without needing to adjust your body.
2. Close your eyes and bring your attention to your breathing. Notice the pattern of your breath without trying to control it.
3. Now, begin breathing deeply, allowing tensions to seep out of your body with each exhale.
4. When you feel more relaxed than you did at the start of this exercise, bring your attention to your feet. Scan every part of each part of them. How do your feet feel? Are you picking up on any sensations? Is there tension? Take a deep breath in, imagining you're breathing light into your feet, and then exhale slowly, releasing all tension and discomfort from them as you do.
5. Move your attention up to your calves and do the same thing you did in step 4. Work your way up your whole body, front and back, until you arrive at the crown of your head. Remember, breathe out the tension in each part.

Deliberate Acceptance
1. Sit somewhere silent, free from devices, distractions, and disturbances.
2. Take a few deep breaths and become aware of the present moment.
3. Now, pay attention to the thoughts and feelings that well up within you.
4. As each thought or feeling floats into your conscious awareness, accept it. Regardless of how boring or bizarre it is, don't attempt to fight or judge it. See them for what they are: fluctuations in your mental field.
5. As you observe your feelings and thoughts, notice how they bob up to the surface of your awareness and dissolve while you remain detached from them. Understand that this process is how you place yourself in a state where you observe quantum probabilities.

Do you get the feeling that this book suddenly took quite a sharp turn? That's deliberate. You have a rudimentary understanding of quantum physics. Now, it's time to take what you've learned so far and see how it all fits in with the mystical, spiritual aspects of life. You're on the train to the Twilight Zone. Buckle up and remain safely seated with your hands, arms, feet, and legs in the vehicle at all times.

Chapter 6: Quantum Mysticism – Science and Spirituality

The connection between quantum mechanics and spirituality may not be immediately obvious, kind of like how you'd never think of mixing ice cream with a bowl of chicken noodle soup!

When it comes to all things quantum, there's a connection between science and spirit, as you'll soon discover in this chapter. The interplay between both fields will offer you some of the most intriguing ways to view reality, your identity, and your role in the grand scheme of life and the universe.

When it comes to all things quantum, there's a connection between science and spirit. Designed by Freepik. https://img.freepik.com/free-photo/mystical-numerology-scene_52683-107763.jpg?t=st=1712100292~exp=1712103892~hmac=cb45e539f861710558dd2229477b9caf3 7f47e6752aa6296bd82769a8bf50c66&w=740

Before the ride begins, know that quantum mysticism isn't something accepted by the entire scientific community. Some say it's an oversimplification of the intricate nature of quantum mechanics at best or a misrepresentation of quantum principles at worst. Nevertheless, quantum mysticism will make you think long and hard and give you some *aha* moments about the weird things you've noticed about your life.

The Ride Begins: Key Aspects of Quantum Mysticism

Some people call quantum mysticism "quantum woo" or "quantum quackery" because they think it's ridiculous. If they took a moment, they'd find their derision is rooted in fear because they're scared of the implications that science and spirituality can find common ground, and they'd be forced to reevaluate their preconceived notions about life and how it works.

Thankfully, Deepak Chopra, Stuart Hameroff, Fritjof Capra, Gary Zukav, Lawrence LeShan, Arthur Koestler, the Fundamental Fysiks Group, and other great minds in the New Age space are unperturbed by pejorative opinions on quantum mysticism and have played their part in bringing it to the forefront of humanity's awareness. They couldn't care less either about Wikipedia's obvious push to make it seem like nothing more than "woo woo" nonsense. Quantum mysticism is the metaphysical bridge that connects quantum physics to consciousness, mysticism, and spirituality. What are the key aspects and ideas of this field?

Non-Locality: According to classical physics, locality is a principle that suggests the only way there could be any physical interaction between two particles or objects is when they're near each other. In other words, the further apart they are, the less likely they are to affect one another.

Quantum physics suggests this classical take on locality leaves much to be desired in explaining spooky action at a distance. It suggests non-locality instead, the idea that it doesn't matter how far apart particles are once they're entangled. They'll always affect one another, regardless of how many galaxies or lightyears you put between them.

Looking at non-locality through the lens of quantum mysticism, it's apparent that everything in the world is intricately connected. If everything is really made up of subatomic systems, whether photons, electrons, quarks, or other particles, there's something that keeps it all connected.

You're connected to the entire cosmos. In fact, you could say you *are* the cosmos. You're part of one great, big, cosmic soup of particles, all connected to one another. Electromagnetism, gravity, and quantum forces all work to connect all galaxies, planets, people, plants, animals, and objects to one another. So, whatever affects the one affects the whole – regardless of how much space or time is between them.

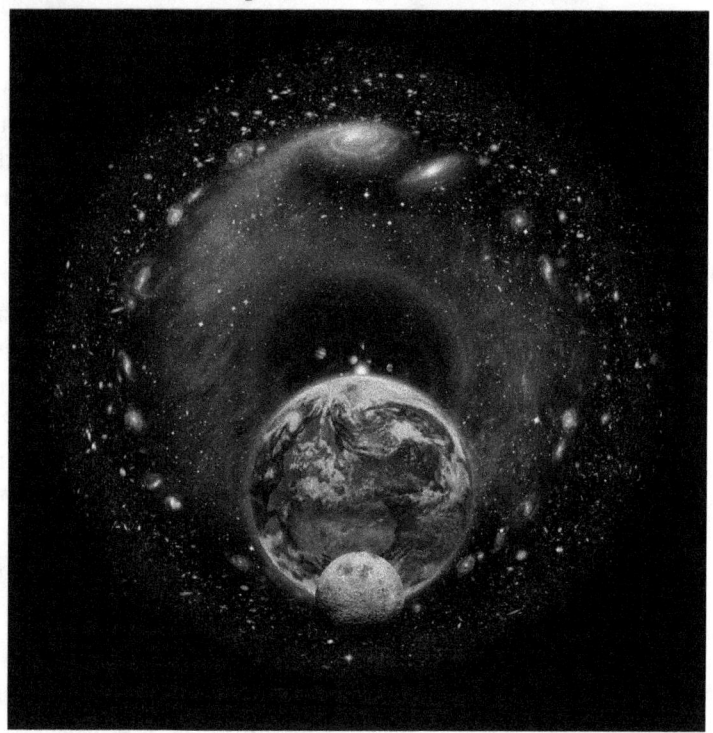

Quantum forces all work to connect all galaxies, planets, people, plants, animals, and objects to one another.

Pablo Carlos Budassi, CC BY-SA 4.0 <https://creativecommons.org/licenses/by-sa/4.0>, via Wikimedia Commons https://upload.wikimedia.org/wikipedia/commons/3/32/Earth_and_Universe.jpg

Interconnectedness and Unity: You already understand the basic quantum premises of quantum physics that suggest all things are connected to one another, such as quantum entanglement. So, what's the connection to quantum mysticism? The Hermetic principle "As above, so below" captures this beautifully well, suggesting that the microcosm and the macrocosm reflect one another.

You are a reflection of the universe, and vice versa. The sacred texts of the Upanishads have the phrase "Tat Tvam Asi," which means "You are That." It essentially is about the fact that everything in existence is one and the same, an expression of divinity.

Buddhism has the idea of pratītyasamutpāda or paṭiccasamuppāda, which implies that all things are caused by something else. Put all of this together, and it's obvious to see how your thoughts, intentions, and emotions affect the world around you – and then some.

Unity of Mind and Matter: The observer effect clearly demonstrates that the observation process affects how reality plays out. Without a doubt, consciousness affects all quantum processes, which affect everything in all worlds, known and unknown.

Buddhists are well aware of this, as they believe in Sunyata, which means "emptiness." What's that about? Every phenomenon is connected to others. This idea is also called dependent origination, where nothing has an intrinsic nature.

Therefore, your life is defined by the phenomena surrounding you, which means your experiences, in turn, are determined by you. What makes a table a table, as opposed to a python? It is the fact that the table depends on other "surrounding" traits that define it as such. Who you are as a person depends on the environment or context you're in.

Your life isn't a permanently defined thing. A scammer could be someone else when they find themselves in a different context. "Context" isn't only the physical environment, but the mental as well. If there's something you don't like about your life or experience, changing your mental context or the way you view yourself is an excellent way to transform yourself into the person you would rather be.

Reality Creation and Manifestation: Put together all the previous key aspects of quantum mysticism, and it becomes obvious that you now have the keys to crafting the life of your dreams. While your life may appear firmly fixed and unchangeable, every particle remains in a state of superposition until you observe it.

Feeling stuck? It's time to take your attention off things as they are. You can do this by practicing mindfulness meditation, as you've learned in the previous chapter, remaining in a state of non-reactive awareness. You know there are infinite versions of you, including the one you desire.

So, from the non-reactive state, you turn your attention to a vision of yourself as you'd like to be. If you're not great at visualizing (another way to say "imagining pictures"), you could focus on the *feeling* you'd have as the person you desire to be. This is how you cause the wave function to collapse into this new, preferred version of reality. You live your life by observing yourself as this new person, thinking, feeling, and acting through them.

Thanks to quantum entanglement, the changes you've put in place in your mind will cause the world around you to adjust as needed, becoming more and more in alignment with the version of yourself you've "collapsed" into. Working with the quantum Zeno effect, you maintain your attention on being this preferred version of yourself, and this sustained attention or observation keeps you in this new reality you've chosen for yourself. This is how the Law of Attraction works.

Unified Field Theory

Have you ever played with Lego blocks? If you have, you know they come in various colors, shapes, and sizes. Yet, despite their differences, they all connect together because they're designed to. The unified field theory, a term coined by Albert Einstein, is an attempt to connect all forces of physics to one another. It's about seeking the one theory to rule them all.

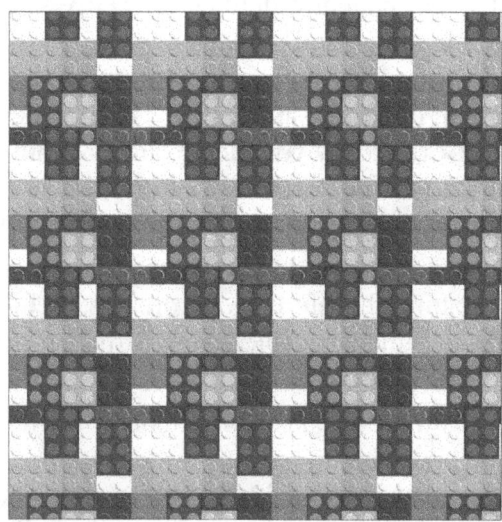

Legos are a good metaphor for how they connect together despite their differences.
https://www.needpix.com/photo/download/1733437/lego-background-lego-building-blocks-pattern-lego-bricks-shape-design-education-toy-pattern

Physics dictates that it's not that these forces are transmitted from one particle or object to another. Rather, there are unique entities known as fields that are responsible for deploying these forces.

The unified field is a field that rules over every aspect of life, connecting the strong nuclear force, weak nuclear force, gravity, and electromagnetism. What this theory implies is that everything is connected at a fundamental level, and if there were a way to flesh out this theory accurately, the implications for physics and the breakthroughs to come would be beyond imagination.

Now, what about the quantum field? This field is what lies at the heart of all physical aspects of reality, containing information and energy in the form of virtual particles with the unique trait of being able to pop in and out of existence. If you want to understand why matter and energy are the way they are, look no further than the quantum field.

The unified field theory has the goal of uniting all of nature's forces, while the quantum field is the glue that holds all energies and particles together, whether you're thinking about subatomic particles or every single universe in existence. This field is responsible for the non-locality that connects one thing to all else.

Quantum mysticism explores the quantum field and unified field through spiritual lenses. Avid meditators in the zero thought or zero point state are simply connecting with these fields, from which they can craft the realities they prefer. The same applies to other mindfulness and spiritual practices, such as yoga. By making a habit of connecting with these fields each day, you'll experience a deep sense of appreciation for life, purpose, passion, joy, and oneness with everyone and everything around you.

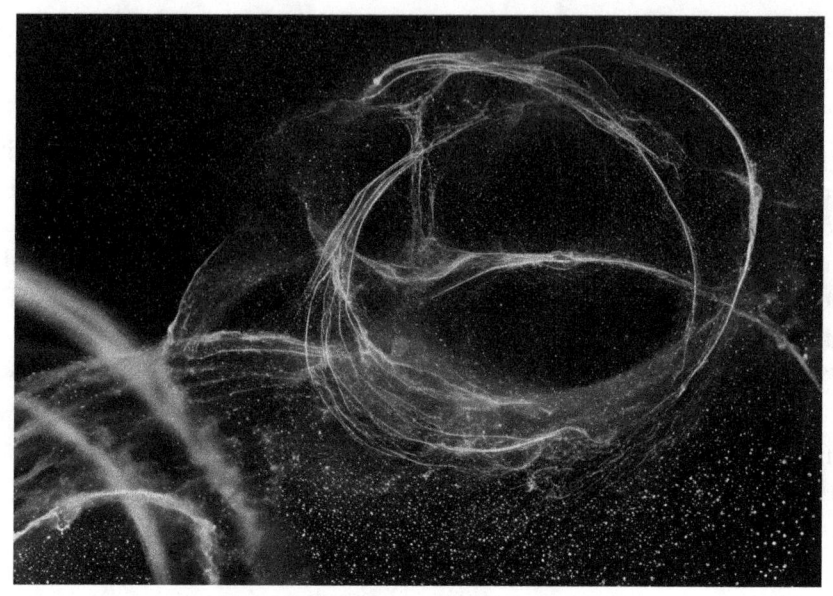

The unified field is the source of all things.
https://www.publicdomainpictures.net/en/view-image.php?image=527494&picture=quantum-physics-waves-and-particles

The unified field is the source of all things. It is where all of life and its phenomena come from and to where it all returns. This field is pure potential, with nothing fixed unless and until you fix your attention on one probability, disregarding the others. The unified field theory suggests that all forces and particles are really the creation of the same energy, something that spiritualists have always known long before quantum physics was a thing.

Another fascinating thing about the unified field theory is the fact that it goes beyond the limitations of duality. It's no longer this or that, but this *and* that, as Bashar, channeled by Darryl Anka, is known to say often during his sessions. All things are one and the same, varying only in frequency or degree of expression. It's like how heat and cold are really expressions of one thing: temperature. This non-duality is reflected in spiritual teachings such as Hinduism's Advaita Vedanta, or Tibetan Buddhism's Dzogchen.

Spiritual teachers such as Deepak Chopra and "the sleeping prophet," Eckhart Tolle, have shed light on the similarities between experiencing the quantum field and enlightenment. If you understand your true nature isn't your name, age, job, looks, or any other ego attachment but is actually the field itself, you'll become aware of who you really are.

You will no longer be swayed by physical reality. You'll know just how plasticky it is and never settle for anything less than you desire. Knowing yourself as and in the quantum field means discovering the true meaning of enlightenment, which is ultimately how to break free from the bonds of suffering, lack, and limitation.

Jung's Synchronicity

If you think coincidences are a thing, think again. There are no accidents in this universe. Still not convinced? That's fine, but can you think of a time when a series of events so implausible occurred, forcing you to pause and wonder if there's not a divine, supernatural force winking at you or having fun at your expense?

Carl Jung was a brilliant mind who came up with the term synchronicity, which demonstrates there's a powerful connection between your mind and the universe. Synchronicity is about "coincidences" pregnant with meaning, at least as far as the observer is concerned.

Carl Jung stated that there is a deep connection between the physical world and your mental world.
https://picryl.com/media/eth-bib-jung-carl-gustav-1875-1961-portrait-portr-14163-cropped-c7875d

These events are a-causal, which is a fancy way of saying when it plays out, you can't point out which event was the cause and which was the effect. It would be like throwing double sixes six times in a row as soon as the clock strikes six while you're wearing a football jersey with the number six emblazoned on it. How many sixes would it take for you to realize there's nothing random about that occurrence in that example?

According to Jung, there's a deep connection between your mental world and the physical one around you. You draw information from the collective unconscious, a cloud or field from which everyone else receives insights, inspiration, vision, revelation, and more.

This collective unconscious is full of archetypes and experiences that are relatable to one and all rather than relegated to an individual's experience. These Jungiang archetypes show up in everyday life in the form of synchronistic events. What's the purpose of these events, you may wonder? Well, they're either delivering you messages or showing you what's going on in your inner, mental world.

Carl Jung didn't have much to do with quantum physics, but his work on synchronicity, upon closer inspection, shares a philosophical connection with quantum phenomena. Think about non-locality, for a moment, as particles that are linked with each other mirror each other, and you'll find how that concept parallels Jungian ideas of the connection between the world and your psyche through entanglement and the observer effect.

Did Jung specifically talk about a connection between synchronicity and the idea that all things are preordained? Not really. It was something he preferred to view with an open mind. At the most, he saw synchronicity as something of a psychic experience that deserves your rapt attention and keen study.

Still, some view synchronicity as more than meaningful coincidences, choosing to see them as messages from a higher power, the universe, or Source Energy, if you prefer. This worldview holds that synchronicity is a sign that the universe isn't the chaotic result of some random big bang but that there must be an intelligence running things, ensuring all of creation remains on a divinely ordered path.

The Non-local Nature of Quantum Phenomena

It's not hard to find the connection between the quantum phenomena of non-locality and spiritual experiences that defy logic and aren't bound by spacetime limitations. You already understand entanglement, but what about Bell's theorem? Here's how that works.

You have a pair of dice. These are no ordinary dice, either. They're magical. Throw them, and they'll always give you the same value, regardless of how apart they are from each other in space or time. Plop one in Timbuktu and the other in Pluto, and they'd still have the same value. Transport one die to the caveman or Anunnaki era and another to the year 5078, and they'd still have the same value. *How is this any different from entanglement?*

Enter Bell's Theorem. Classical physics suggests that at some point when you throw both dice, they should have different values because non-locality isn't a logical expectation or phenomenon in macroscopic physics. Yet, when it comes to entangled particles (or, in this case, entangled dice), they'll always match each other – with no regard for the limits of classical physics.

Therefore, Bell's Theorem declares that there's no compatibility between quantum mechanics and a theory involving local hidden variables. Put simply, if the universe only functions the way classical physics dictates, then that means spooky action at a distance is impossible; but as you already know, it's not only possible but proven.

This chapter is about quantum mysticism, so it's best to circle back to the topic. Bell's Theorem and quantum entanglement have parallels with transcendent spiritual experiences. Have you ever had a moment when you felt connected to the world around you to a profound, inexplicable degree? What about a moment when it felt like time no longer existed?

During special occurrences such as these, you lose your ego. You lose awareness of everything in your mind that has you thinking you're separate from the world outside yourself. You merge with the cosmic soup of energy, becoming one with all. Moments like these show you there's a dimension of life beyond linear time. Your soul understands the first few lines of William Blake's poem, Auguries of Innocence:

To See a World in a Grain of Sand

And a Heaven in a Wild Flower

Hold Infinity in the palm of your hand

And Eternity in an hour

Not only do you lose your ego and all sense of time, but you also develop truly unconditional love and acceptance for one and all, even those you never thought you could feel compassion for. So, what do these spiritual experiences have in common with non-locality in quantum physics? They're both phenomena that take place in a realm where clocks and maps are irrelevant.

Both also reflect the interconnectedness of all life. They also demonstrate that there's something even more real than what your five senses pick up on in this physical world, which you cannot grasp, at least not with your present understanding of the universe.

Exercises to Tap into the Unified Field

The takeaway from this chapter is there's value in maintaining your sense of unified awareness. The question is, how can you pull this off? Using the following exercise, you'll transform your life for the better through the fastest and most efficient means by shifting your experiences from the quantum level rather than through effort and fruitless action.

That's not to say action has no place in changing your life, but once you attain that unified awareness, you'll find the actions you take are less like an uphill battle and more like rowing your boat merrily downstream like life is just a dream – a lucid one that's very responsive to your thoughts, feelings, and intentions.

Being here, now: Find a comfy position to sit or lie in, and then close your eyes. Direct your attention to your breath. Part your lips slightly. Take a deep inhale through your nostrils, hold the breath in for a few seconds, and then exhale through your mouth. It's natural for your exhale to be longer than your inhale, so don't overthink it. Simply notice your breath. Give it a second or two before you repeat the process once more.

You could pay attention to the sound of your breathing, the feeling of the air as it goes in through your nostrils and out through your lips, the gentle rise and fall of your chest and belly, or the counts for each portion of this awareness exercise.

Do this for ten to fifteen minutes each day. You'll find it useful to have a timer set before you begin so you don't keep distracting yourself by checking to see how many minutes go by. *Let the timer do the worrying on your behalf!*

Be warned: Your mind will wander away from your breath. It could do this as many as three times every 45 seconds or a hundred times in a minute. When you notice this, it's imperative that you don't beat yourself up for losing focus. If anything, that's worth celebrating; you're learning to notice when your mind wanders!

So, gently and lovingly release the distracting thought or feeling, and return to your breathing as many times as you notice you've been distracted. With time, you'll notice you're less and less distracted.

The benefit of being mindful of the here and now through this deliberate meditation will reveal itself to you in the coming days and weeks as long as you're consistent. You'll discover the power you have within to change your world, and not only that, you'll be less reactive to the things that triggered you into undesirable states of fear, anger, and anxiety.

From this more empowered state, you can envision the life you want, hold your goal steady in your mind, and breeze through the process of transformation as life shifts from what you don't want to what you prefer.

Chapter 7: Entanglement – Everything Is Connected

In this chapter, you'll explore the spiritual dimensions of life. Not only that, you'll learn more about the interconnectedness of all things, as demonstrated by quantum entanglement through the lens of spirit. So, are you ready to go even deeper into the rabbit hole? Good. You're going to love it!

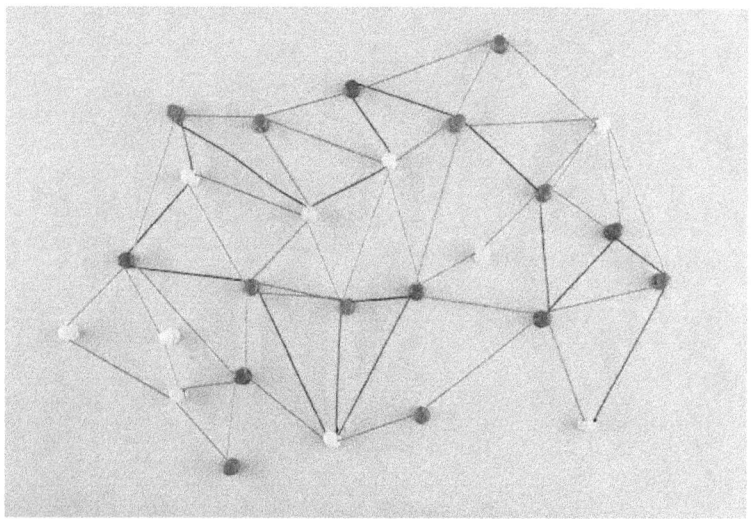

Everything is connected.
Designed by Freepik. https://www.freepik.com/free-photo/network-concept-with-colorful-thread_15292480.htm#fromView=search&page=1&position=1&uuid=9da01092-4a93-48ce-9ba7-196234a14a3e

Entanglement: A Metaphor for Spiritual Interconnectedness

There are other implications of quantum entanglement outside of science. Where else can you apply the idea of two particles, separated by space, still sharing a connection so strong as to mirror each other? You'll find the theme of interconnectedness across religions – echoed in spiritual beliefs and practices. It's not simply about being at one with others and your world but being at one with the very source of all creation.

All of life is a cocreation if you really think about it. Everyone and everything has a part to play to keep life going. When you think about the observer effect and the fact that at every point in time, there are billions of points in space observing everything else, you'll realize how truly interconnected everything is.

Suddenly, the butterfly effect doesn't appear so outrageous. Every observer is doing their part to shape reality as it is, thanks to consciousness. Your intention and attention matter as much as the next person's, dovetailing perfectly together to create outcomes that match expectations, regardless of what spiritual path or practice you use to create your desired reality.

Look at the world around you, and it would appear that duality is the order of the day. There's "us versus them," "black versus white," "up versus down," etc. It never ends, or so it seems. Quantum entanglement and spirituality suggest otherwise. All duality springs from the unity of consciousness.

Remember, entanglement is about a particle's ability to reflect another one it is entangled with, which means everything, whether good or bad, is simply a reflection of the collective consciousness of humanity. It's all connected, much like Indra's web in Buddhism.

Are you familiar with Reiki? It's a spiritual healing modality that involves restoring balance to the body and mind on an energetic level. Quantum physics holds that the universe is all consciousness or, if you prefer, energy and information. The particles that make up the world are all part of this field of energy.

Practices like Reiki tap into specific healing energies from the body's bioenergetic field.
alfonso.saborido, ATTRIBUTION 2.0 GENERIC, CC BY 2.0
<https://creativecommons.org/licenses/by/2.0/>
https://www.flickr.com/photos/28063292@N02/22560032539

In spiritual traditions and practices like Reiki, the point is to work with this field to achieve your goals since everything in the field is connected to everything else. It's all fields within fields. Your body has its bioenergetic field, and since it's connected to the unified field, it makes sense for Reiki practitioners to draw specific healing energies from that field into yours to help you heal.

Unity in Diversity

Quantum entanglement is proof of unity in diversity. On the surface, this sentence may seem paradoxical. After all, the very essence of diversity suggests that the parts are distinct and separate from each other. Otherwise, how could you tell them apart from one another?

However, when you view life through the lens of quantum entanglement or with spiritual eyes, it becomes blatantly obvious that everything is really one and the same, regardless of how separate they may appear to be to the ordinary senses. This is not a call for you to lose your sense of self or to assume that no one or nothing is special. Your uniqueness is as valid as the fact that you are united with the world around you in energy and spirit.

Hinduism is one of those spiritual paths that emphasizes the idea of unity in diversity. If you follow this path, then you believe that Brahman is the ultimate reality that does not have any specific form. The Brahman is also infinite and beyond time, for it is both primordial and eternal. While it is the very essence from which all of life is crafted, it is also represented by the Hindu pantheon of deities, which represent its different divine aspects.

For instance, there is Shiva, who is the God of destruction and regeneration. Saraswati is the embodiment of knowledge. Lakshmi is the Goddess who is the essence of wealth and prosperity. These are just a few of the Hindu pantheon of gods and goddesses. While Hindus respect and revere each one of these divine beings because of their unique qualities, they are seen as being part of the Brahman. In this way, Hinduism reflects the truth of quantum entanglement, which is unity in diversity.

What about the Sufis? What do they believe? Sufism is a mystical aspect of Islam. One of the tenets this philosophy espouses is the idea that all creation springs forth from the divine and contains an element of divinity within it. This is known as the unity of being or Wahdat al-Wujud.

A true Sufi will tell you that there is nothing and no one in existence that does not contain the divine creator's essence within them. All of creation is a manifestation of Allah's divinity. It is impossible for anything or anyone to exist without the will of Allah keeping them in existence.

Sufis believe in a higher order for everything to connect for existence.
Peter Morgan, ATTRIBUTION-NONCOMMERCIAL-NODERIVS 2.0 GENERIC,
<https://creativecommons.org/licenses/by-nc-nd/2.0/>
https://www.flickr.com/photos/pmorgan/3189477502

Have you ever heard of Indra's net? Indra is a Vedic Deva. Hanging over his palace on Mount Meru, the net has a jewel in each node. An interesting truth about these jewels is that they reflect one another. If you think about it, this is a perfect representation of the interconnectedness of everything in the world. It is a beautiful way to visualize unity in diversity.

Buddhism also has the concept of *Pratitya Samutpada*, also known as dependent origination, which emphasizes that each phenomenon, known and unknown, is not only connected to the others but also exists because of said others.

Have you ever heard of Yin and Yang? No, not the rapping twins. Yin and Yang is the Taoist concept of opposites harmonizing with each other. While these forces are clearly distinct from each other, they depend on each other to keep the world performing in balance as it should.

Imagine a world where there's only up and no down or left and no right. That would be a rather strange world to live in, wouldn't it? Yin and Yang express the idea of unity and diversity wonderfully well by taking polar opposites that make up duality and blending them together to create a harmonious existence.

Society has brainwashed people into demonizing one end of the spectrum over the other. Extreme conservatives don't want to hear out extreme liberals because the other side is full of demons or deluded people. This is a sad way to live life because even a broken clock gets it right at some point during the day.

Then there's the argument over what's more superior, the masculine or the feminine. In a world that does not recognize or respect unity in diversity, you get Andrew Tate at one extreme and Shera Seven at the other. Being unable to find the balance between light and dark is a recipe for disaster, and in case you wondered, "dark" in this context is not a bad thing. That thought process would be akin to saying the night and moon are evil and the day and sun are good.

There will always be those who argue that everything is one, and that's all there is to it. Then you have those who argue the opposite, claiming that it's ridiculous at best to suggest everyone's one and the same and invalidating at worst. This is the lovely thing about quantum entanglement, as it acts as a bridge between both philosophies. It demonstrates that the universe, while fundamentally united, is a space that allows diversity and uniqueness to thrive. The whole is not greater than the sum of its parts, and vice versa, as one can't exist without the other.

Meditation and Contemplation: Bridges to Quantum Power

How do you take advantage of quantum entanglement? If two particles are interconnected on a quantum level, affecting each other instantaneously, it stands to reason that you could achieve the same thing with your life. Think of yourself as a particle and your experience of life as another particle with which you are intricately connected.

Up until this point, you may have allowed the external reflection of your life experience to dictate your state of being. However, since you are entangled with your life experience, what if you simply switch your state of being without waiting for the world outside of you to do so first?

For instance, if you want more happiness and rewarding relationships in your life, what if rather than waiting for people to show up who are a perfect fit for your desires, you embody the state of being a person who already has these fulfilling connections?

According to the principle of quantum entanglement, your life will have to reflect this brand-new state of being that you have adopted. The most effective way to take advantage of this power is to become deeply aware of your connection with the world around you. It must sink from basic head knowledge into your heart or feeling center and head down into your belly: *you are not separate from everyone else.*

When it's "in your belly," you live your life consciously aware of the other parts of yourself. It's easy to assume that simply accepting this truth in your mind is enough, but that's not the case. You need to really get this down in your gut. How do you accomplish this? Through the mechanisms of meditation and contemplation.

Meditation makes your experience of oneness and unity very real and palpable. You can't meditate for three hours straight once every four business months and assume that should get the job done. It's much better to be consistent with your practice. Even just five minutes a day will help you become more aware of your connection to all things, and with time, you'll carry that awareness even after you've finished sitting in silence.

Many practitioners of meditation report experiencing a point when it feels like they expand beyond or collapse into their bodies, becoming everything or nothing. You get the sense that there's no way to tell where you end and the world begins. Words can only do so much to capture the profound experience of unity that comes through meditation. You have to put in the work and see for yourself.

Contemplation is another powerful tool similar to meditation. When you meditate, you deliberately release your thoughts and feelings, observing them with no attachment. However, when you contemplate, you take time to ponder specific topics. The goal of contemplation is to receive deep, profound insight and understanding regarding your spiritual walk.

By giving your mind the task of unpacking the idea of quantum entanglement, you reveal your manifestation power to yourself. Quantum entanglement suggests that all things are interconnected, which would imply that if you want something, you already have it since you are that which you desire.

Remember: *Tat tvam asi.* "You are that." The seeker is the sought. Awaken to the truth that you contain all you want and need, and you no longer have to seek these things out. Your desires will find you. This is just one of many realizations you arrive at when you practice contemplation.

Quantum Meditation

Quantum meditation is no ordinary form of meditation because it involves working with quantum physics principles. The way to work with this meditation modality is to accept that your thoughts, feelings, and intentions have real effects on your experience of life. The process of thinking and feeling interacts with the quantum field.

Remember, this field is full of potential, which means that your thoughts and feelings are your observations of the field. Your observations, in turn, crystallize a specific, matching outcome out of the many probabilities the field offers you. So meditation not only helps you with manifestation but also reminds you that you are eternally connected to the quantum field, making it easier for you to manifest your heart's desires.

How do you incorporate the principles of quantum physics in a practical way as you meditate? Consider the observer effect and you'll understand the importance of using your imagination to visualize the preferred version of yourself you prefer. You will also be working with the power of intention to supercharge your visualization.

Intention is your will. It is knowing that not only is your preferred reality possible but also a done deal. With this attitude, you will influence your physical reality so it mirrors the visions that you have in your mind of how it should be.

A powerful goal of quantum meditation is to enable you to feel oneness with the universe. More often than not, when people meditate, it's because they're looking for ways to alleviate feelings of anxiety, depression, worry, etc. People meditate to find inner peace.

However, quantum meditators seek much more than stillness. They want to experience the life that they know they deserve to live.

Even when they don't have a specific desire they'd like to bring into this 3D reality, quantum meditators continue their practice to remind themselves of the interconnectedness of all things and remain in the awareness of their unity with "the all," or the unified field, if you prefer.

Quantum meditation is a combination of mindfulness and quantum physics principles that lead you to a state of superconsciousness. Here's how to use quantum meditation to accomplish whatever your heart desires.

Make a decision. You must know exactly what you want from life. Far too many people are well-versed in what they don't want. In fact, you may have answered the question, "What do you want?" by rapidly reeling off everything you can think of that you'd like to stop or end or wish wasn't a problem for you. If this is the case, you're focused on the wrong end of the stick.

To make this more practical, you shouldn't be saying you want a better-paying job than the one you have when what you really want is more money for less work. If you're not clear about what you want yet, you can use the things you don't want to give you clues about what you *do* want. Then, go a step further by asking yourself why you want those things, and you'll discover your true desires.

For instance, you think you want a lot of money, but when you drill further, you realize what you really want is to travel the world. You assumed you'd need a lot of money to pull off exploring the world's beautiful treasures when that isn't true. What if all your flights, accommodations, meals, and miscellaneous needs were sorted out by someone else? So, get specific about what you want, and you'll get out of your own way.

Get comfortable. Find somewhere quiet, free from distractions and disturbances, where you can focus for the next 10 to 15 minutes. Make sure you're dressed comfortably so your clothes don't feel like they're itchy, too tight, too warm, etc.

Sit comfortably. Do you have a recliner? That's perfect. Set it to a semi-upright position. If you don't have one, it's fine. Sit on a chair or *on* the floor on a mat in a lotus or half-lotus position.

Close your eyes and bring your attention to your breath. Take a few deep inhales and exhales, focusing on releasing all tension and worry as you exhale. Keep breathing like this until you notice you're feeling calm and still, fully present in the here and now.

Imagine. This is the same thing as visualization. Picture yourself in a quantum zone. You can make it look like whatever you want it to. It could be a white or black void or the beach. It could even be a hallway full of doors that branch off to different versions of your life.

Whatever you visualize, know that this zone is where everything and anything is possible. This is the zone where every possible version of you exists. There are no limits here to the choices you could make. You can select various timelines and parallel realities. Ensure you are imagining from a first-person perspective rather than third-person. In other words, you shouldn't be seeing your body as separate from you. You should be within your body.

Imagine that multiple versions of reality branch out from your present one. Whatever you do, don't be in a hurry to choose the closest or quickest. Instead, select the path that you desire the most. This means you'll have to check logic at the door. The quantum zone is beyond logic and rationality. You'd be doing yourself a great disservice by binding yourself to these things.

Choose. Pick the version of your life that calls to you the loudest, the one that feels right in your soul. The process of choosing could look however you want it to. It could be you walking through a door, portal, or liminal space of some sort. It could look like flicking through channels on a television to find the version of reality you prefer and then stepping through the screen to embody that life. The choice is yours.

Use your imaginal senses. Once you step into the version of life that you prefer, use your 5 senses in your imagination to make everything feel real to you. What can you see in this scene? What do you hear? What can you smell, taste, and touch? How do you feel emotionally? The deeper you immerse yourself in your imaginal senses, the more real it will feel to you and the more you charge this new version of your life, forcing it to become your new normal.

Accept it is done. Resist the temptation to dismiss this exercise as "nothing more than imagination" when you've finished. Imagination is a tool that allows you to interact with the quantum field and pull from it whatever you desire. Those teachers and other adults who reprimanded you for daydreaming as a child owe you a million apologies.

There is an important point to remember if you decide to practice quantum meditation to manifest your dreams. Whatever you do, you should imagine the different options available to you *long after* you have received your desire.

If you want to get a car, you don't imagine yourself at the dealership trying different vehicles on for size. Instead, you imagine yourself, say, six days, weeks, months, or a year after you've got the vehicle. This way, you fix it firmly in your mind that your desire is no longer a desire – but is the reality of your situation right now. As the great mystic Neville Goddard put it, you're making "there" here and "then" now.

You can use this quantum meditation technique for whatever you want. You could use it for relationships, friendships, promotions, and healing. Did you have a terrible night? Were you unable to sleep long enough? You can use this meditation to place yourself in a version of reality where you got the best night's sleep and you'll be amazed at how effectively this works.

Working with the quantum field through this form of meditation changes the neurochemical workings of your brain, putting you in touch with the power to choose rather than roll over and accept the hand fate has dealt you. Through quantum meditation, you'll come to recognize your interconnectedness to the world and become more aware of your consciousness. This is how you heal your life. This is how you manifest your dreams.

Chapter 8: Superposition: Anything Is Possible

Superposition is quite a paradox, isn't it? Imagine particles are like ice cream flavors. How interesting would it be to have the same scoop be every flavor imaginable at the same time? In this chapter, you'll dive deeper into the idea of superposition to understand how it works even better and find the connection between this quantum physics phenomenon and spiritual concepts like the power of focused attention and intention.

Superposition: Unlimited Potential

Schrödinger's cat thought experiment is the epitome of superposition. Remember, it's about a quantum system being able to exist in more states than one until the observer effect kicks in. This is still a bit difficult to accept because it would imply that the red car in your driveway is also orange, yellow, blue, green, purple, upside down, broken down, brand new, and also *not* in your driveway unless you're looking at it. So, what does superposition have to do with spirituality?

Regardless of the spiritual path you consider, which culture it's from, or from which period of humanity's history it originated, you'll find there's a belief that every person carries a spark of divinity within, which gives them value. Spirituality suggests everyone has inherent worth, thanks to this *divine spark*, as the Christians call it.

In Buddhism, this spark is *Buddha nature*. In Hinduism, it's the *Atman*, a word to describe the true self. In Islam, it's the *fitra*, the part of human nature that honors the *tawhid* or oneness of God. This part of you is primordial purity. By embracing these spiritual tenets, you learn that it makes no sense to pigeonhole yourself with labels. These are concepts that show you that you're far more than your physical self.

You have the potential to be, do, or have all that you can imagine and then some, but it's all going to remain dormant until you choose to express that potential within you, whether spiritually, emotionally, intellectually, or in any other way.

Superposition in quantum mysticism invites you to move from black-and-white thinking to play more in the gray. Ditch the limitations of "either/or" and embrace "this *and* that." This way, you'll become aware of the various abilities, skills, talents, and experiences available to you and choose what you want from that smorgasbord for a fuller, richer life.

If you've lived long enough, you've probably come to accept that life will always have ups and downs, good and bad, highs and lows. All of these things are ultimately good because you evolve and grow by experiencing both sides of the spectrum. In the darkness, you discover new aspects of yourself, just as you do when it's light.

As a result, you become aware of what's possible for you. When this happens, you can't go back to being who you used to be without feeling miserable and unfulfilled. It's best to stretch and grow beyond that. As the process of observing a super-positioned particle forces a collapse in its wave function, it assumes a specific state, and so do the experiences in your life compel you to embody and express the part of you that was once no more than a dream.

Many people find themselves up in arms about what they're supposed to do with themselves. What's the point of life? Why go on? These, and many more, are some of the existential questions that humanity's forced to ask of itself. How do you know you're supposed to be a pilot instead of a pirate? How do you know you're meant to be a matchmaker rather than a manager? Well, you don't. You're not "supposed" to do anything other than explore yourself for the rest of your life. That's it.

Self-exploration and self-expression are the ultimate reasons for living. You can't fight the thing within you asking you to grow in one direction or the other. Try as you might, change is inevitable. When you make peace with exploring beyond your comfort zone, you'll learn more about who you are. Your true self is full of wonderful, sometimes mind-bending surprises if only you keep an open mind and maintain your affinity for falling down rabbit holes. Your inspired thoughts and ideas are in a superposition state and will remain there until you act on them. Only then can you bring them into reality and see how you like it.

The concept of manifestation is a popular one in spirituality. You could think of manifestation as the process of bringing forth your quantum desires into the physical world to experience them. Until your desire becomes a reality, it remains in a state of superposition.

So, you can look at the quantum phenomenon of superposition as a metaphor for the potential that you carry within you, which is yet untapped and unlimited. Your desires remain in the quantum realm unless you choose to manifest them and make them real in the physical plane. Manifestation is the ultimate form of creativity.

You now understand what superposition is about. How do you take advantage of it? How do you put it to work? To create the life of your dreams, the first thing you must do is accept that everything is possible. Just because you do not see a path to the preferred outcome you desire right now does not mean there isn't one or that you can't have what you want.

In the same way that the quantum particle exists in a myriad of states simultaneously, you also have a myriad of ideas that exist within you.

There is a version of you with that car, house, significant other, or whatever else you seek. There is a version of you that is healthier than the current version you're embodying. There is a version of you that lives a fulfilled life and has finally found true love. Whoever you wish you were, you are that person, but you are being that person *potentially*, in a state of superposition.

The way to force a wave function collapse and become this potential person you'd like to actualize is by first accepting that everything is possible so you don't limit yourself to the things you are already familiar with. Don't let logic act like a ball and chain, keeping you from spreading your wings and flying.

The next thing you must do is visualize the outcome you prefer. By developing a mental picture of how you'd feel, think, and go about your day as this person that you want to become, you force the wave function collapse to occur. You force the superposition state of your potential into becoming crystal and firm in one state only. Once you have this clear mental snapshot of yourself, you should take action. Action is another important part of the process of collapsing your superpositioned ideas and preferences into reality.

Intention and the Potential for Personal Transformation

When you contemplate the idea of superposition, you realize there's no such thing as a fixed reality. The truth about reality is it's dynamic. There are countless possibilities that play with one another to create even more interesting possibilities. If you would like to transform your life completely, then you need to give some thought to what your conscious intentions are because they are the propellants of the change you seek.

Due to the limitless nature of the quantum realm, the change you could experience could absolutely rock your world. Limitations are only as real as you think they are. For far too long, humanity has gone about assuming that it is impossible to change reality. This assumption is further entrenched in the human psyche thanks to organizations and systems with rules and processes that must be followed to the letter.

The seeming rigidity of the nature of reality has gone unquestioned for long enough. You could think of it as a blessing that quantum physicists have discovered and continued to research the idea of superposition, which suggests that reality is as fluid as can be. It is full of infinite possibilities that continue to flow and evolve as they interact with each other in an ebb and flow.

Looking at this idea through a spiritual lens, you'll find that reality is a matter of creation, of interconnected threads that are ever in a state of flux, responding to whatever you're thinking or feeling in the moment or whatever intention you fixed in your mind.

If you desire transformation like nothing you've experienced in your life, you need to use the power of intention. At this point, the logical question to ask is, *what exactly is intention?* What does it mean? Some people think of intention as nothing more than goal setting. They think it's only about making plans and attempting to follow those plans.

An intention is much more than that. It is what gives life to transformation. It is a prayer, silent and sacred, that you whisper with sincerity to the universe or your Creator, trusting that it will be expressed as your reality.

When you express your intention, you aren't simply uttering words for the sake of them. You're bringing every part of you into play. You're infusing these words with energy or feeling.

Your intention is the one thing that you live for. It's what you wish to experience above every other thing in life. More often than not, intentions are tied to those things that you don't consider possibilities, in the sense that you may not think of the concept of intention unless and until you notice you've been struggling with a particular goal for a while. But why is that the case? To understand the answer to that question, you must think about the nature of an intention.

Conscious Intention

For one thing, intentions are clear. Are you one of those people who consistently makes plans only to have them fall through, and you've decided that planning is an exercise in futility? It could be because you did not bring clarity into the mix from the get-go. You must be clear about what you intend to experience or accomplish first. This means tuning into what your deepest desires are.

Once you become clear, you must feed your intention with faith, which is an inherent trust that this thing you desire not only can manifest but is as good as done. To have faith is to go beyond believing to knowing that it is done. From this state of knowing you have your desire, you can then take action.

When you combine clarity, faith, and action together, you have a powerful intention that must grow and become *that which you seek in reality*. This is the secret, not-so-secret formula to creating your reality with the power of intention and transforming your life in ways beyond imagination. Your intention acts as the observer collapsing the wave function into your desired outcome.

Whether your intention is big or small is insignificant. In fact, the ideas of big and small are nothing more than logical in positions that you place upon yourself. As Abraham, a collection of entities channeled by Esther Hicks, often says, "It is as easy to create a castle as it is a button."

You would be doing yourself a huge favor by doing away with those assumptions that some things are more difficult or will take longer time than others to manifest. All you need to do is remain consistent with your intention by keeping your attention focused on what you desire and acting in alignment with that desire, assuming that you already have what you want. By doing this, your intention will blossom into an actual manifestation.

Some people understand the idea of intention and work with it to manifest their desires, but they fail. Why does that happen? There's a key piece of the puzzle that's missing, which, once you have it, will unlock the doors to the impossible for you forever. This missing piece is repetition.

Those who attempt to manifest their desires and get no results often assume that once is enough. It's possible to get to the point where things happen that fast, but when they're just starting out with learning to manifest and don't have enough faith, it's not a good idea to only work with intention now and then. If this is you, there's no reason to beat yourself up for not knowing. Repetition is powerful.

When you repeatedly restate your intentions and focus on them, you cause all your thoughts and actions to line up with the preferred version of reality that you seek. Nature abhors a vacuum. If you are thinking, acting, feeling, and living like someone who already has what you want, you are causing a vacuum, and therefore, nature must swoop in to correct that vacuum by giving you the life you're acting as if you already have.

This is quantum entanglement in action, where whatever's happening to one particle must happen to the other particle it's entangled with. Repetition is how you learn everything, isn't it? It's how you became so proficient at reading and writing. Well, the same process is how you become proficient at living your life as this new version of yourself that you are as yet unaccustomed to being. Think of it like watering a plant and applying fertilizer so that when it blooms, it blooms beautifully.

Visualization

When you visualize, you create a powerful, clear picture in your mind of what you prefer to experience in your life. Visualization is an excellent tool to catalyze your personal growth and encourage the manifestation of your dreams. There appears to be a vast chasm between your desires, which are in a state of superposition, and the manifestation of said desires. Visualization is the bridge that connects these two together. As you picture yourself, your world, and your life being the way you'd prefer, you cause a wave function collapse.

Visualization is like selecting a specific channel on which to watch a specific show. For instance, say you'd like to watch something by the actor Ryan Reynolds. You've seen him in a plethora of comedic shows and movies, but you'd like to see a more serious side of him. So you scroll through all the options of everything he's ever been in, and you finally choose the one thing where he plays a serious character.

You know this is different because you can see a visual representation of Ryan Reynolds being serious versus being goofy. When it comes to manifesting your desires, you are Ryan Reynolds in this context. You are also the person with the remote control who gets to choose which show you'd like to see. You accomplish this using visualization. Visualization is powerful when it is repeated. Each time, you keep your focus on the version of yourself you would prefer to be.

As you visualize, you should never imagine yourself being projected on a screen; instead, embody yourself by seeing through your own eyes. Some people have used visualization and found it doesn't work for them, but for others because whenever they practice visualization, someone else winds up with their manifestation.

If you practice visualization by looking at yourself as if you're on a screen, you are projecting your desires onto someone else. But by embodying yourself, looking through your eyes, and being inside your body as you visualize your preferred outcome, you ensure that your manifestation is yours and yours alone.

Affirmations

Affirmations are statements made in the present sense to attest to the fact that you already have whatever it is you want. Superposition offers the perfect metaphor to understand how affirmations can crystallize into actual manifestations in your life. Every word that you speak is a seed that must bear fruit sooner or later.

The Bible says life and death are in the power of the tongue. While that may sound like an utterly dramatic statement, it is not far from the truth because often, as the Bible also says, out of the abundance of the heart, the mouth speaks. Whatever you truly believe about yourself, your life, and others is exactly what you'll say unless you're deliberately trying to deceive someone or actively changing your life through the power of your words.

A key part of affirmations is repetition. As you repeat these affirmations, you cause your subconscious mind to buy into them more and more each day. You are causing the wave function collapse that will convert your present reality into the desired one. As you repeatedly affirm your preferred truths, you'll discover that your actions and thoughts are in alignment with these new statements.

You're missing out if you don't take advantage of the power of affirmations because there's no better way to change your belief systems. Now the question is, why would you want to change what you believe? The answer is simple. You can't manifest what you don't believe. An excellent definition of a belief, according to Abraham Hicks, is it's a thought that you've been thinking over and over long enough that you now think it's the truth.

A key part of that definition is the idea of repetition. If you have installed beliefs in your mind that do not serve you and will not help you accomplish the dreams that you want to see become real, you would be doing yourself a favor by working with affirmations that support your new preferred life. If you'd like a different experience, then you have to install a new belief system, and there's no better way to accomplish that than by using repeated affirmations consistently.

The moment you adopt new beliefs is when life will shift for you. This is because your beliefs act as a filter. So if you believe that life is full of hostile and terrible people, the kindest person in the world could pass by you on the street, smile at you, and say hello, and you would somehow find a way to misinterpret that greeting as being malicious. Once you believe differently that life is full of wonderful, genuine, kind people, you'll begin noticing more of that in your life because you have a new filter that supports a life full of pleasant people around you.

Now that you understand the power of conscious intention, visualization, and affirmation, here is an entry-level process that incorporates these tools to manifest your reality:

- **Set clear** and precise descriptions of what you desire.
- **Set an intention** framed in the present tense based on your desire. Keep it short and simple.
- **Relax.** Close your eyes, and get comfortable. Breathe deeply until you are aware of only the present moment, and then imagine yourself doing, being, or having whatever you desire.

- **Repeat positive affirmations** that speak to the truth of what you desire as being real. Do this at least once a day for five to ten minutes at a time, either at the start or at the end of your day.

Your affirmation is your intention in words. Keep each one short and simple, and always start with the words "I am." If your conscious mind keeps butting in with logic, telling you that you aren't who you say you are, you can use "ask affirmations" instead. How?

Ask yourself questions like, "How did I become so wealthy?" "How did I become so healthy?" You're not asking these questions to get actual answers. You're only asking them the same way you'd ask your significant other, "How did I get so lucky to be with you?" Whether you're using **ask**firmations or **aff**irmations, repeat them over and over with feeling and gratitude.

- **Act in alignment with your intention to the best of your abilities.** Eventually, ideas will spring up within you about which course of action to take next. Follow every hunch you receive. Act with the awareness that it's already done, and even if your action does not prove fruitful, assume that it's already done.

Just because you walk into a movie theater while the protagonist is down and out doesn't necessarily mean that's how the movie ends. If you've already seen the movie before, you're not bothered by that one sad frame. You know the good guys come out on top in the end. This is the same attitude you should approach your experiences as you set about the business of manifesting your dreams.

It should offer you some comfort to know that there is a version of yourself who already has everything you could ever dream of and has an entirely different set of desires and goals that they'd like to accomplish. This version exists not only because of the quantum phenomenon of superposition but also the fact that the multiverse is a sound theory. Not even Zuckerberg's metaverse, with all his smarts and dollars poured into it, could ever hold a candle to the supernatural multiverse.

Chapter 9: The Multiverse

According to the Many-Worlds interpretation of quantum mechanics, the multiverse is real. Every quantum event could lead to a myriad of outcomes, and each of these outcomes causes a separate branch of reality to be formed. So, in this final chapter of the book, you are going to delve deeper into the multiverse. You'll understand the ins and outs of this theory and see how it can reshape your assumptions of life for the better.

The multiverse.
https://pixabay.com/photos/parallel-world-parallel-universe-3488497/

The Multiverse

According to the hypothesis of the multiverse, the world is full of multiple universes besides the one in which you currently live. Some of these universes may closely mirror what you're familiar with, while others could be so far removed from anything that you've ever known or imagined.

The multiverse theory suggests there's a world in which gravity works in reverse; you breathe in carbon dioxide to breathe out oxygen, and the digestion process works from the bottom to the top. That last example is, granted, a little disgusting, but that's the multiverse for you. There are no limitations, and there's no such thing as "impossible" because this is a quantum theory with principles that suggest it's a valid one, while yet unproven.

There are four possible kinds of multiverses that you could experience.

1. The inflationary multiverse.
2. The quilted multiverse.
3. The quantum multiverse.
4. The brane multiverse.

The Inflationary Multiverse: Do you remember The Big Bang Theory? No, not the show, the actual theory that the birth of the universe was the result of a Big Bang. After this phenomenon, the whole universe began to inflate like a balloon or a bubble. Scientists say that ever since the Big Bang, the universe has been expanding outward.

If the inflationary multiverse is a thing, this begs the question, is every universe out there also in the process of expansion? According to the theory, the multiverse is a field of energy. This field is infinite and ever-expanding, full of universes that are also in the process of expanding. For this to be possible, the field of energy obviously must exist beyond the limitations of space and time. Scientists also suggest that each of the bubbles or universes may have their own unique laws of physics.

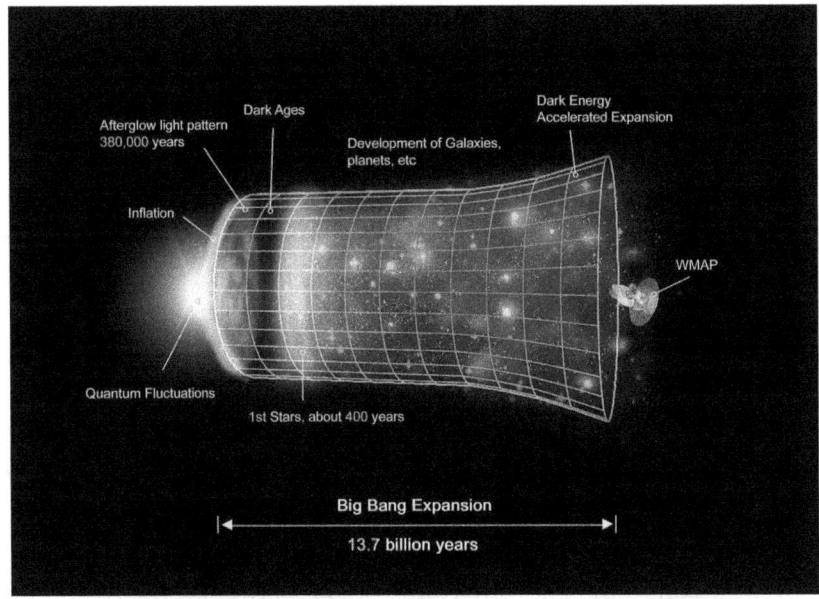

The Multiverse.

The Quilted Multiverse: This multiverse is actually one universe, but it's infinite. The question is, what is the degree of its infinity? Scientists suggest that if you were somehow to develop the means to explore the infinity of this quilted multiverse, you would find a galaxy with a planet that has someone exactly like you on it, doing exactly what you are doing right now.

The space within the quilted multiverse is so infinite that it can house all kinds of probabilities, from the ones closest to your reality to the vastly dissimilar ones. Before you ask, yes, there's definitely a "Potterverse" and a Game of Thrones universe according to this take on the multiverse.

If this is true, why is it that (for now) it seems impossible for you to make your way outside of your zone of existence? The universe is ever-expanding in every direction. Also, there's nothing that travels faster than light. Even if you had always traveled at the same speed of light since the start of your existence in your universe, you'd never be able to travel outside of your world.

The Quantum Multiverse: The quantum world, as you already know, is a world full of quirks and strangeness. The laws of classical physics are so skewed that quantum physics had to be developed to explain the strange phenomena observed at this level of existence.

The Quantum Multiverse.

You're already familiar with some of these strange happenings like quantum entanglement, superposition, the observer effect, etc. This is a multiverse that is highly influenced by your observations, choices, and intentions.

The Brane Multiverse: Conjure a three-dimensional book in your mind. Now, see that book as having two-dimensional pages. According to this view of the multiverse, your universe is only one of these pages in your imaginary book.

The Brane Multiverse.

Now imagine that same book, but make it 10-dimensional this time. What's really happening, according to this theory, is that the universe exists on something called a brane or a membrane. This universe is an intrinsic part of all 10 dimensions of your imaginary book.

What about the other pages of your book? Well, those are other universes. According to this theory, it's possible to have black holes within the pages that act as portals that can send you to another page or universe. Some say these portals can transport you to another "book" entirely, with its own infinite pages of universes and black holes.

What about bubble universes? This is a unique model. Your universe is really in a false vacuum, which is a state of energy that's not stable. It's possible for pockets of "true" vacuum, which have much lower energy, to form bubble universes within your regular, physical world, where the laws of physics don't line up with the classical laws. It's possible to never stumble upon these pockets of wonky reality, let alone communicate with any life within it.

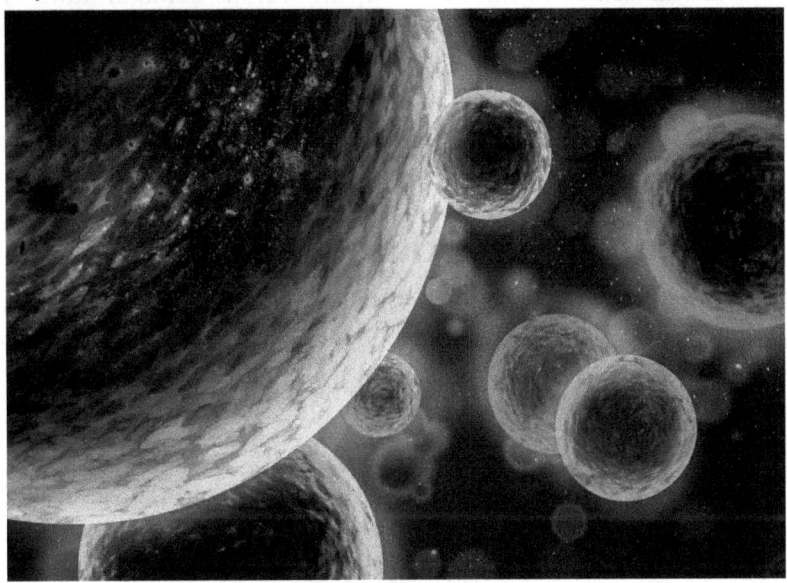

The Bubble Universe.

String Theory: Extra Dimensions

According to string theory, it is quite possible that there are other dimensions– aside from the ones you already know – that exist in the three-dimensional world in which you live. It implies other dimensions that do not adhere to time as you know it either. String theory proposes that each of these universes has different laws of physics specifically tied to the nature of these unknown and unfamiliar dimensions. It is thanks to these extra dimensions that scientists can figure out certain inconsistencies that pop up in physics and also combine gravity with other fundamental forces in a cohesive manner.

One interesting thing about these extra dimensions, according to string theory, is that they appear to be curled up on microscopic scales infinitesimally smaller than an atom. According to scientists, it is this compact size that explains why you don't observe these extra dimensions in your daily life.

These dimensions aren't curled up randomly either but are in specific sizes and shapes. You can think of little circles or complex-looking shapes. Sacred geometry is a part of this topic, but it is beyond the scope of this book. However, that's a subject worth diving into when you have the time.

The extra dimensions can be set up in various configurations. Each configuration is a unique one that distinctly matches a special universe with its own constants and variables. The number of potential universes you could find in the string theory landscape is astronomical and unimaginable. String theory is the womb from which the multiverse emerges.

Implications of the Multiverse

What if the multiverse is really real? What are the implications for humanity? How will that affect the way people think of reality, consciousness, and the concept of free will? The process of pondering the idea of the multiverse, let alone accepting it, opens up a Pandora's box of unimaginable implications and questions.

If everyone other than quantum physicists were to pause and reflect on this for longer than 5 to 10 minutes a day, it would probably cause the economy to come to a crashing halt. At the very least, it would force many people to question the point of their existence on this planet.

One of the most challenging and immediately observable implications of the multiverse theory is that if the multiverse is a real thing, it would suggest that so many different life forms are yet to be discovered and probably never will be.

It would appear that aliens are a thing. It would also suggest that there's no such thing as a fixed history since every single quantum event leads to countless opportunities that branch off into new universes. For too long, people have viewed the universe through anthropocentric lenses. But what if there are actually alternative realities besides the one about which you are so knowledgeable?

Now, move on to the topic of consciousness. Is consciousness a truly universal phenomenon, or is that something that's localized? According to the many worlds' interpretations of the multiverse and other interpretations, there are an infinite number of copies of yourself in parallel worlds. They all have their unique struggles and traits.

This begs the question, what does it mean to be yourself versus your other selves? Also, is it possible that, as depicted in the Oversoul Trilogy books by Jane Roberts, there is a universal consciousness that underlies all these versions of yourself? What if you aren't the whole story or person? What if you're only a part of a larger, grander version of yourself, much like one cell in your body does not make up your entire biological system? Can you see how this makes the problem of consciousness even harder?

Following from the previous paragraph, you have to wonder whether or not free will is real. Is your entire life already predetermined? In that case, what's the point in making plans and trying to see them through? Or do you actually have choices? Think about it. If it turns out that your universe is only one out of countless universes that already have outcomes that are set in stone, then does free will actually exist?

Do you truly have the power of choice? Some great minds suggest that if the multiverse is real, this could actually add to your free will rather than take away from it. They suggest if there are multiple versions of yourself and they're all selecting different paths of life, that can only give you even more choices as your potential increases, and you have a wider range of experiences that, even if you haven't actualized them yet, remain accessible to you in their super-positioned, potential form.

Spirituality and Science: Bridging the Gap

The multiverse isn't the brainchild of quantum physics. Humanity has contemplated this idea for millennia. Science is only beginning the journey of proving the existence of this multiverse theory. For now, all you can rely on is the subjective evidence of the multiverse that is offered to you through spirituality.

There are countless tales of people who have successfully shifted from one reality to another and have subjective evidence. Unfortunately, science is a field that scoffs at anything anecdotal. Hopefully, scientists will find something objective that proves the multiverse is indeed real. For now, you will have to make peace with exploring how ancient philosophers, cultures, and traditions viewed the idea of many worlds.

Anaximander was a Greek philosopher from the 6th century BCE who considered the idea of an infinite "apeiron," from which everything in existence springs. Granted, his thought was not necessarily about the multiverse, but it did confirm that there might be some other reality that births this physical one.

There is also the idea of the One and the Many. According to Plato's Theory of Forms, there is a world of forms that remain unchanging because they are perfect. Forms serve as molds or archetypes from which the world as you know it was created – and is still being created. He referred to this perfect world of forms as "the One" and the imperfect world, where you exist and perceive the creations of the One as "the Many." This is yet another idea from ancient Greece that suggests there is more than one reality.

Make your way over to India and study Vedic cosmology. Comb through ancient Indian texts and analyze the Vedas, and you'll notice that there are descriptions of universes that are cyclical in nature. According to these texts, these universes create themselves only to dissolve and then be recreated. If you think about it, this almost mirrors the idea of there being countless universes with their unique traits. Hindus believe that there is one true reality underlying all of existence. This reality is known as unity. It is the Brahman that gives birth to the physical world as you know it. The physical world, according to the Hindus, is the world of Maya or illusions.

In the Chinese concept of Dao, there is a formless principle that encompasses all of existence. This principle is akin to the unified field theory of quantum physics, where everything exists in a state of potential, waiting for a waveform collapse to become one thing rather than remain all things. Traditional Chinese philosophy recognizes that there are multiple realms besides the physical world, spiritual and otherwise.

What about indigenous and shamanistic beliefs? Those who adhere to this way of life understand that there are other worlds besides this physical one. In fact, they journey to these other worlds using shamanistic practices. Some people have out-of-body experiences and lucid dreams where they travel to these mystical places. Others make use of psychedelics or other substances to get them there.

Still others employ different modalities, such as singing, dancing, chanting, and drumming, to get to these alternate universes, often returning with useful insight, revelations, guidance, and more for those who need such assistance.

It is important to include shamanistic and indigenous opinions and experiences in this book because, despite all the other theories postulated, shamans and similar people have actual evidence of other realms, even if their experiences are subjective. There are so many tales of people entering into different realms through dreams or the aforementioned modalities only to return with critical life-saving (or life-changing) information. These are well documented, so it boggles the mind why scientific society remains so brazenly dismissive of these accounts to date.

There truly is only one way to prove to yourself that there's more to life than meets the eye. What's that? You need to have your experience of these so-called subjective, nonexistent, yet unproven dimensions of existence. You could do this in many ways, but among the safest methods are lucid dreaming and out-of-body experiences.

If you have no idea where to start, you should definitely check out Robert Monroe's series on astral projection, where he shared his experiences in great detail, offering you some perspective on what to expect when you begin your journeys. The wonderful thing about Monroe's work is that he conducted his experiments and research in a scientific manner, so you can finally shut the inner cynic within you and take the plunge to discover what lies beyond the veil.

When you finally prove to yourself that there are worlds beyond the one you are familiar with, you'll open up your mind to the thought that there is indeed a version of you who already is exactly the person you would like to be. You'll also see how you can become what they are.

Remember, the multiple versions of yourself are not unlike quantum leap entangled particles. By focusing your attention on what you hope to accomplish in life and seeing yourself as having already done that, you will mirror the version of yourself who has already attained success. Your life will have no choice but to reflect to you the inner work or inner shift you've accomplished as a result.

Conclusion

At the start of this book, you were promised a lot of weirdness. And you must admit, every page delivered on that promise. Quantum physics is the most counterintuitive thing you'll ever encounter – other than spirituality, of course. Every phenomenon in theory in this field of study is absolutely mind-boggling and forces you to pause and reflect on what you think you know about your life. Whatever you do, don't let your exploration of quantum physics end with this book. The deeper you dive into the topic, the more you will experience paradigm shifts that will benefit you in every way you can imagine.

Before you move on to reading something else, you should take time to reflect on what you've learned from these pages so far. Consider the fact that this whole time, you may have assumed that you are nothing more than a passive observer of your life. Perhaps, like many others, for the longest time, you thought that you had no control or say over how your days should go. You figured you had to eat whatever slop was served on your plate, not knowing that you could treat yourself to an entire buffet if you wanted it. With your newfound knowledge of the observer effect, you no longer have to stand by as a passive witness in your life.

You no longer have to settle. If you want something better for yourself, you absolutely can go for it because you now know that you are a conscious co-creator with the universe. You now understand that if the universe is indeed predetermined, it is at least predetermined by your observation, intention, and will. You now understand how constantly observing the same old thing gives you more of the same old thing. No longer will you allow life and its many vicissitudes to walk all over you. Instead, you take your place as a god of sorts and dictate how your life goes.

With your knowledge of superposition, you break free from the classic black-and-white thinking. As a result, you put yourself in a state where you can experience quantum shifts. According to classical physics, you should go from the first gear to the second, the third, and then the fourth. However, according to quantum physics, you can simply skip from the first to the umpteenth.

Too many people are held back in life by their rigid thinking, by their assumption that everything must happen in a logical sequence, and that it is impossible for a thing to play out in any other way than what is conventionally known and accepted. You are no longer among this class of people. You have now been set free. You recognize you have access to a world of infinite possibilities, and you will milk that opportunity to live a rich, fulfilled life for all its worth.

If you take the time to contemplate what you've learned from this book, you'll discover that there is power in focusing your attention on whatever you desire. You'll learn not to be swayed or dissuaded by the physical reality which appears to be in contrast with your desires.

Your confidence in your ability to get what you desire or even better will come from the fact that you now know all you have to do is keep your intention sure and strong – and continue to act in line with it. Above everything else, you'll recognize the artificial and unnecessary nature of the divisions that keep people from realizing that, at the end of the day, everyone is made of "star stuff," and everyone's one and the same, imbued with the power of the creator of all things.

Here's another book by Mari Silva that you might like

Your Free Gift
(only available for a limited time)

Thanks for getting this book! If you want to learn more about various spirituality topics, then join Mari Silva's community and get a free guided meditation MP3 for awakening your third eye. This guided meditation mp3 is designed to open and strengthen ones third eye so you can experience a higher state of consciousness. Simply visit the link below the image to get started.

https://spiritualityspot.com/meditation

Or, Scan the QR code!

References

Banik, M., Gazi, Md. R., Ghosh, S., & Kar, G. (2013). Degree of Complementarity Determines the Nonlocality in Quantum Mechanics. Physical Review A, 87(5). https://doi.org/10.1103/physreva.87.052125

Buhrman, H., Cleve, R., Massar, S., & de Wolf, R. (2010). Nonlocality and Communication Complexity. Reviews of Modern Physics, 82(1), 665-698. https://doi.org/10.1103/revmodphys.82.665

Clegg, B. (2009). The God Effect: Quantum Entanglement, Science's Strangest Phenomenon. St. Martin's Griffin.

Dyson, F. (2013). Is a Graviton Detectable? International Journal of Modern Physics A, 28(25), 1330041. https://doi.org/10.1142/s0217751x1330041x

Filk, T., & Albrecht von Müller. (2009). Quantum Physics and Consciousness: The Quest for a Common Conceptual Foundation. Mind and Matter, 7(1).

Hayes, L. J. (1997). Understanding Mysticism. The Psychological Record, 47(4), 573-596. https://doi.org/10.1007/bf03395247

Hirshfeld, A. C. (2000). BOOK REVIEW: String Theory. Volume I: An Introduction to the Bosonic String. by Joseph Polchinski. String Theory. Volume II: Superstring Theory And Beyond. by Joseph Polchinski. General Relativity and Gravitation, 32(11), 2235-2237. https://doi.org/10.1023/a:1001959811458

Horgan, J. (2004). Rational Mysticism. HMH.

Jackson, G. (2022, September 28). What is the Main Difference between Classical Physics and Quantum Physics? [Fact Checked!]. Physics Network. https://physics-network.org/what-is-the-main-difference-between-classical-physics-and-quantum-physics/

Kenneth William Ford. (2011). 101 Quantum Questions: What You Need To Know About The World You Can't See. Harvard University Press.

MacIsaac, T. (2018). A New Theory of Consciousness: The Mind Exists as a Field Connected to the Brain - Science and Nonduality (SAND). Science and Nonduality (SAND). https://scienceandnonduality.com/article/a-new-theory-of-consciousness-the-mind-exists-as-a-field-connected-to-the-brain/

Mansuripur, M. (2009). Classical Optics and its Applications. In Cambridge University Press (2nd ed.). Cambridge University Press. https://www.cambridge.org/core/books/classical-optics-and-its-applications/7E0D316A0E283CAE3876B7DAC50621B4

Misra, B., & Sudarshan, E. C. G. (1977). The Zeno's Paradox in Quantum Theory. JMP, 18(4), 756-763. https://doi.org/10.1063/1.523304

Nomura, Y., Poirier, B., & Terning, J. (2018). Quantum Physics, Mini Black Holes, and the Multiverse: Debunking Common Misconceptions in Theoretical Physics. Springer International Publishing.

Oppenheim, J., & Wehner, S. (2010). The Uncertainty Principle Determines the Nonlocality of Quantum Mechanics. Science, 330(6007), 1072-1074. https://doi.org/10.1126/science.1192065

Ponte, D., & Schäfer, L. (2013). Carl Gustav Jung, Quantum Physics and the Spiritual Mind: A Mystical Vision of the Twenty-First Century. Behavioral Sciences, 3(4), 601-618. https://doi.org/10.3390/bs3040601

Popescu, S. (2014). Nonlocality Beyond Quantum Mechanics. Nature Physics, 10(4), 264-270. https://doi.org/10.1038/nphys2916

Posner, M. I. (1994). Attention: the Mechanisms of Consciousness. Proceedings of the National Academy of Sciences, 91(16), 7398-7403. https://doi.org/10.1073/pnas.91.16.7398

Pratt, D. (2007). Consciousness, Causality, and Quantum Physics. NeuroQuantology, 1(1). https://doi.org/10.14704/nq.2003.1.1.5

Qian, X.-F., Vamivakas, A., & Eberly, J. (2017). Emerging Connections: Quantum and Classical Optics The blurring of the classical-quantum boundary points to new directions in optics. https://arxiv.org/ftp/arxiv/papers/1712/1712.10040.pdf

Rae, A. I. M. (2013). Quantum physics, Illusion or Reality? Cambridge University Press.

Rogalski, M. S., & Palmer, S. B. (1999). Quantum Physics. Gordon And Breach Science Publishers.

Silverman, M. P. (2008). Quantum Superposition. Springer Science & Business Media.

Simon, C. (2019). Can Quantum Physics Help Solve the Hard Problem of Consciousness? Journal of Consciousness Studies, 26(5, 6).

Stapp, H. P. (1999). Attention, Intention, and Will in Quantum Physics. Journal of Consciousness Studies, 6(8-9).
https://www.ingentaconnect.com/content/imp/jcs/1999/00000006/f0020008/971

Tricycle. (2020). What is Dependent Origination? Buddhism for Beginners.
https://tricycle.org/beginners/buddhism/dependent-origination/

Zeilinger, A. (1999). Experiment and the Foundations of Quantum Physics. Reviews of Modern Physics, 71(2), S288–S297.
https://doi.org/10.1103/revmodphys.71.s288

www.ingramcontent.com/pod-product-compliance
Lightning Source LLC
Chambersburg PA
CBHW051846160426
43209CB00006B/1179